P9-APM-782

Surf, Sand, and Stone

How Waves, Earthquakes, and Other Forces Shape the Southern California Coast

Keith Heyer Meldahl

UNIVERSITY OF CALIFORNIA PRESS

University of California Press, one of the most distinguished university presses in the United States, enriches lives around the world by advancing scholarship in the humanities, social sciences, and natural sciences. Its activities are supported by the UC Press Foundation and by philanthropic contributions from individuals and institutions. For more information, visit www.ucpress.edu.

University of California Press
Oakland, California

Library of Congress Cataloging-in-Publication Data

Meldahl, Keith Heyer, author.
Surf, sand, and stone : how waves, earthquakes, and other forces shape the Southern California coast / Keith Heyer Meldahl.
p. cm.
Includes bibliographical references and index.
ISBN 978-0-520-28004-5 (cloth: alk. paper)
ISBN 978-0-520-96185-2 (ebook)
1. Coast changes—California, Southern.
2. Geomorphology—California, Southern. I. Title.
GB458.8.M45 2015
551.45'7097949-dc23 2014048460

Printed in China

24 23 22 21 20 19 18 17 16 15
10 9 8 7 6 5 4 3 2 1

The paper used in this publication meets the minimum requirements of ANSI/NISO Z39.48–1992 (R 2002) (*Permanence of Paper*).

For Susan

Within my hollow hand,
While round the earth careens,
I hold a single grain of sand,
And wonder what it means.

—Robert William Service

Contents

Author Note *xiii*

1 Time, Faults, and Moving Plates: A Recipe for Southern California *1*
2 Tsunamis *25*
3 Earthquakes *43*
4 Disassembling Southern California *61*
5 Waves and Surfing *89*
6 Beaches and Coastal Bluffs *119*
7 Sea-Level Changes and the Ice Ages *143*

Afterword *171*
Acknowledgments *177*
Appendix: Seeing for Yourself *179*
Glossary *195*
Notes on Sources *203*
Bibliography *209*
Index *217*

Geography and Faults

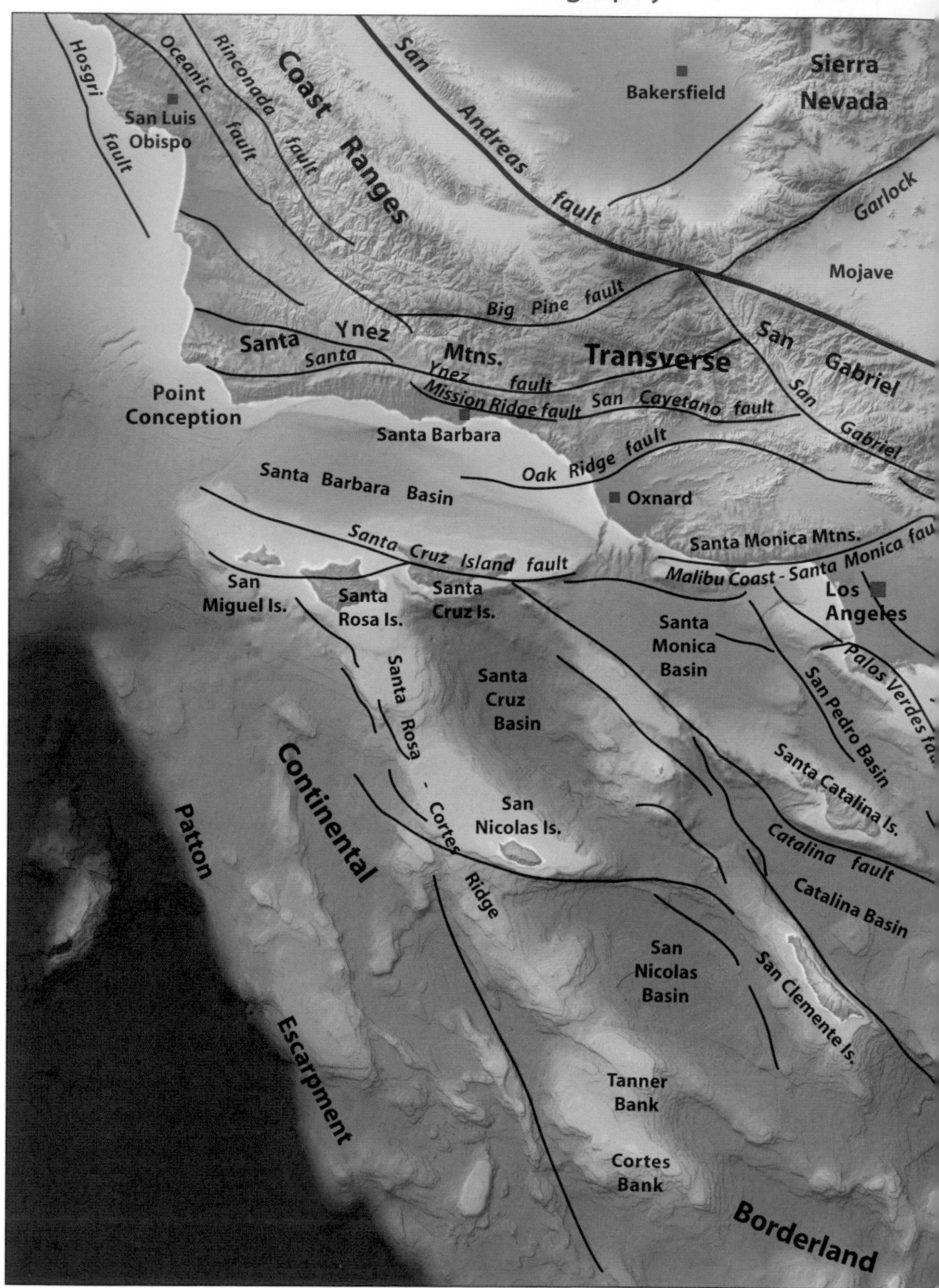

Major geographic features of Southern California, including mountain ranges, islands, undersea basins, and significant earthquake faults. Although this book is focused mostly on the coastal zone, our explorations will take us east to the San Andreas fault and west across the Continental Borderland.

of Southern California

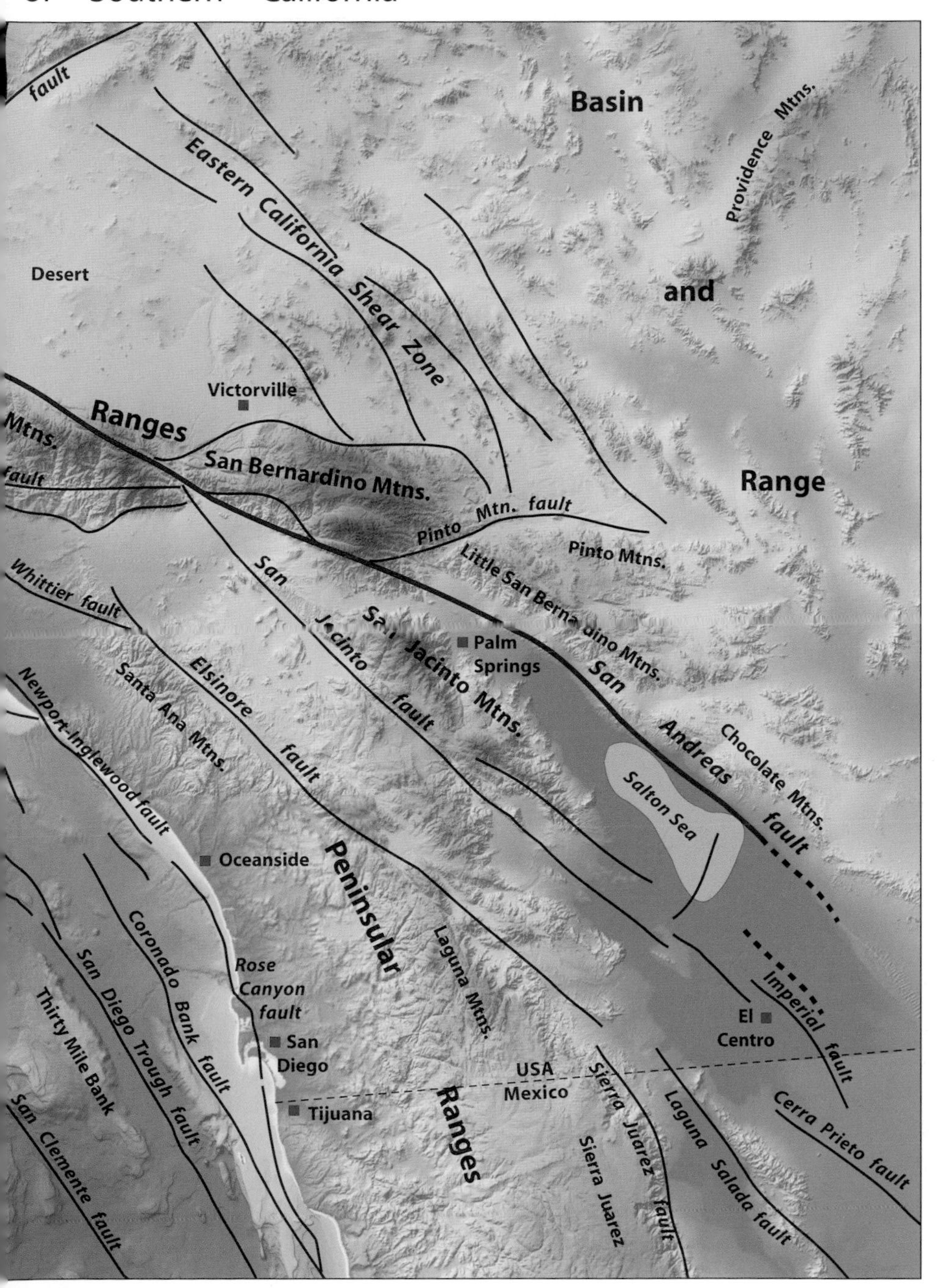

The Earth's

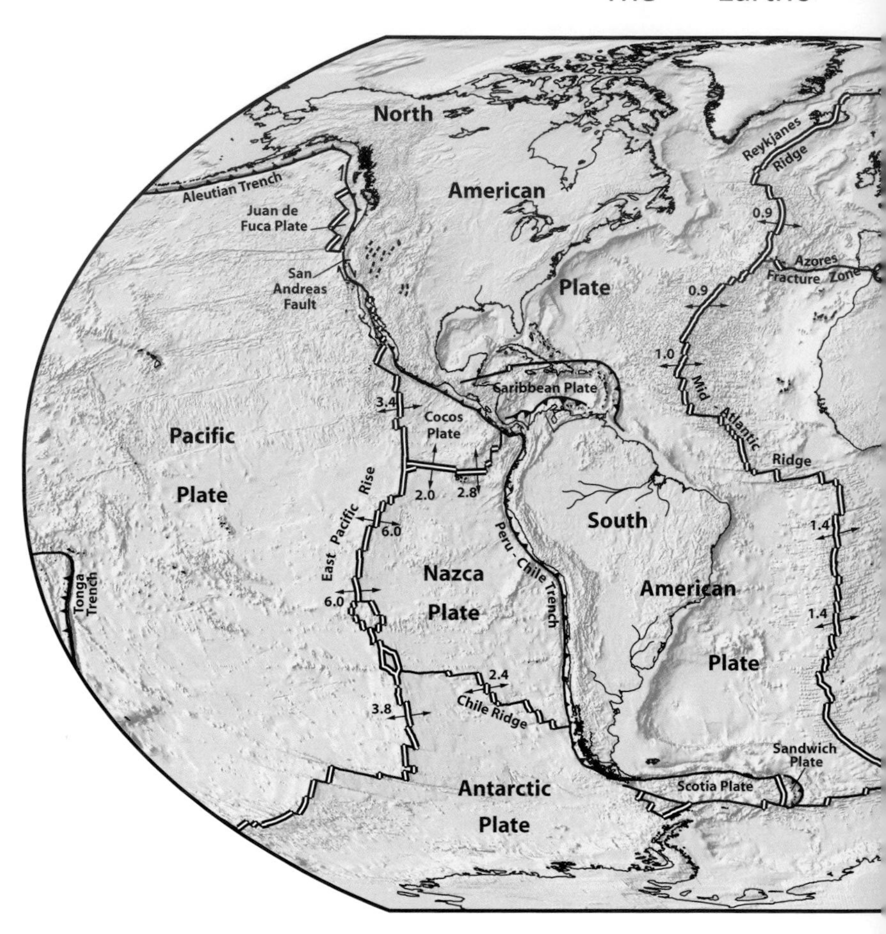

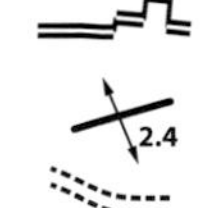

Mid-ocean ridges (where the sea floor spreads) and transform faults (faults that link offset segments of ridges)

Total spreading rate in inches per year

Major continental rift zones

Tectonic Plates

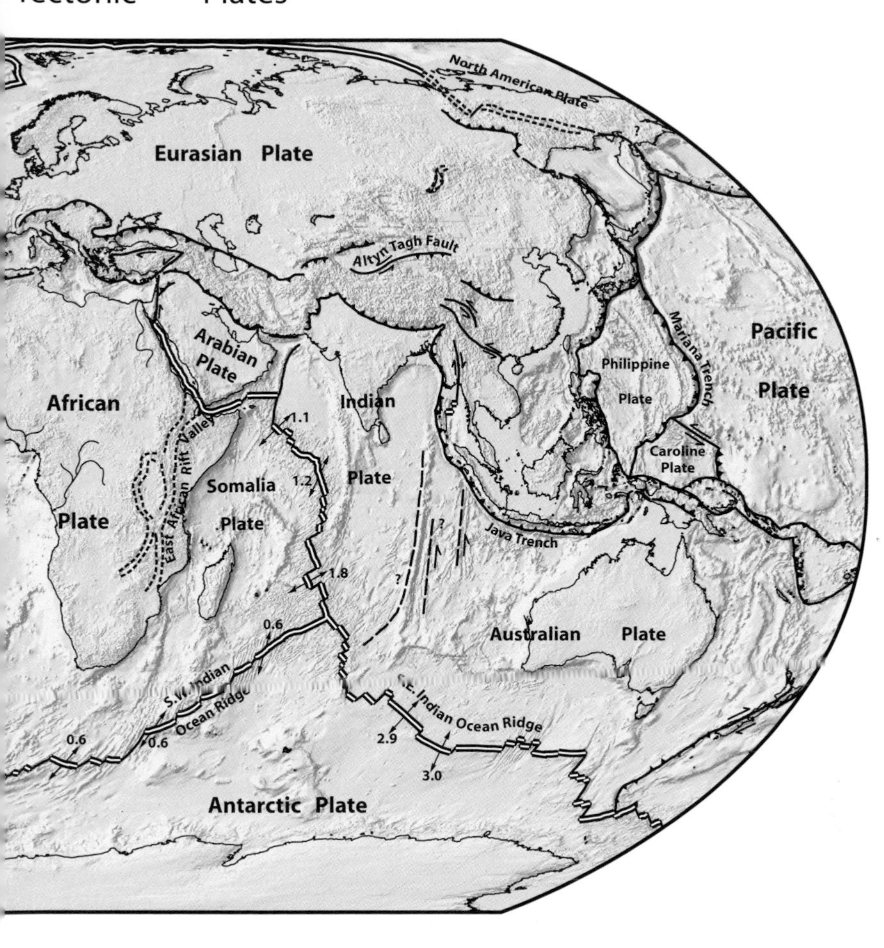

Major transform faults (side-by-side moving faults) with arrows showing sense of motion

Ocean trenches and continental collision zones; barbs show the direction that one plate is sliding under the other

Plate boundary uncertain

Geologic Time Scale

EON	ERA	PERIOD	EPOCH	AGE
Phanerozoic Eon	Cenozoic	Quaternary	Holocene + Pleistocene	2.6
		Neogene	Pliocene	5.3
			Miocene	23
		Paleogene	Oligocene	34
			Eocene	56
			Paleocene	66
	Mesozoic	Cretaceous		146
		Jurassic		202
		Triassic		251
	Paleozoic	Permian		299
		Pennsylvanian		318
		Mississippian		359
		Devonian		416
		Silurian		444
		Ordovician		488
		Cambrian		542
Proterozoic Eon				2500
Archean Eon				3850
Hadean Eon				4550

Ages in millions of years before present, based on the latest Geological Society of America Geologic Time Scale. Note that the vertical scale varies; older geologic intervals cover greater time spans than younger intervals.

Author Note

This book tells the scientific story of the Southern California coast: the story of its mountains, islands, beaches, bluffs, surfing waves, earthquake faults, and other natural phenomena. I tell it by taking you places. That's how I work as a geologist, and for my students, it's a lot more fun than being cooped up in a classroom. You may know some of the places we'll visit: San Miguel and Santa Catalina Island, the Santa Ynez and Santa Monica mountains, the seaport at Long Beach, the terraced mesas of Palos Verdes and San Diego, the surfing waves at Malibu Point and Black's Beach, and many others. I hope that you'll come to know these places in a new way.

This book is about outdoor science. It's about the adventure and fun of exploring one of the most scientifically intriguing places on Earth with the sun on your back and the wind in your hair. Although you can learn everything here from your favorite armchair, you'll have more fun if you bring the book with you to the islands, beaches, and mountains. If sand grains fall from the binding and saltwater stains the pages—or crystallizes on the face of your tablet—you'll know you're using it right.

1

Time, Faults, and Moving Plates

A Recipe for Southern California

Time is Nature's way of keeping everything from happening at once.

—John Wheeler

In one second, the backyard of an oceanfront home may disappear as the bluff beneath it collapses. In one day, a storm may sweep away a beach. In one year, some Californians will see cracks open in their lawns and driveways as their property oozes downhill on slow-moving landslides. In any given decade, the odds are good that a big earthquake will shake California. In fifteen million years, the slow creep of the Earth's tectonic plates will put Los Angeles next to San Francisco. The processes that shape our world operate over a vast range of time scales, from seconds to millions of years. But our lives encompass events on the short end of Nature's clock, and that makes it hard to appreciate the power of long-term change. To understand how the Southern California coast came to be, we need to move beyond human time and get our minds around geologic time, or what geologists call *deep time*. The reason is simple. Geological processes such as erosion, the uplift of mountains, or the movements of the Earth's tectonic plates may seem trivially slow over human time. But give these processes millions of years, and they can accomplish stunning work.

We all know big numbers when we see them, but the years of deep time—millions and billions of years—are hard to grasp. Big numbers are *always* hard to grasp. Who among us, for example, has a good feel for *one billion* of anything? One billion dollars is close to Bill Gates's annual income, so here's a way to put such a number into perspective: one billion dollars per year ÷ 365 days per year ÷ 24 hours per day ÷ 60 minutes per hour ÷ 60 seconds per minute = $31.71 per second. In other words, Bill Gates earns about $32 per second, around the clock, all year long. Now imagine that Gates is walking down the sidewalk when someone stops him and says, "Bill, I need some financial advice. I'll pay you for whatever your time is worth." Gates responds, "OK, that'll be $96 for the last three seconds. How else can I help you?"

If that scenario gave you a better sense of one billion, let's try for 4.56 billion—the age of the Earth in years. Imagine a ninety-five-gallon bathtub (a large household bathtub) filled to the brim with medium-grained sand. (Geologists classify sand precisely, so "medium-grained" means that the grains range from 0.25 to 0.5 millimeters in diameter, which is typical of many beach sands, as well as standard table salt.) Let each grain represent one year. Wet one fingertip and dip it in the brim-full tub. About five hundred grains cling to your fingertip, representing roughly the number of years since Columbus crossed the Atlantic. Scoop up one-eighth of a teaspoon, or about eight thousand grains. That represents the years since the dawn of human agriculture. Now scoop up a heaping tablespoon, roughly two hundred thousand grains. You hold the entire existence of our species, *Homo sapiens*. The brim-full bathtub holds about 4.5 billion grains, representing the age of the Earth. Pour that tablespoon back. Do you see a difference? Compared to deep time, human time is virtually nonexistent.

What does the vastness of geologic time mean for the formation of California? You might think I'm about to make an argument for California's ancientness, but no. Geologically speaking, California is young—although still almost unfathomably old by human standards. Two hundred million years ago (about four gallon-buckets of sand in the tub), California didn't yet exist. North America ended in what is now western Nevada, and had you stood then, say, where Reno is today, you would have gazed west not at the Sierra Nevada and the rest of California, but at ocean waves and deep blue sea. Had you watched for the next hundred million years, you would have seen California arrive, piece by piece, from the ancient Pacific Ocean. Islands, seamounts, and vast chunks of ocean floor, carried by the Earth's tectonic plates, landed on the continent's edge, one behind the other, to assemble California. (I'll give you a fuller explanation of how this happened—and how we know—in chapter 4.) About twenty million years ago, mighty faults, some of them precursors of the modern San Andreas fault, began to slice up this collection of imported rock and send it on the move yet again. That interval—the past twenty million years—is my focus in this book. That's not much compared to the age of the Earth. In the bathtub analogy, it's about seven cups of sand. But it's still a vast span—enough deep time to pack a wallop.

Here's a story to show you how.

MEXICAN PEBBLES FAR FROM HOME

In the ocean thirty miles south of Santa Barbara lie the four Northern Channel Islands— San Miguel, Santa Rosa, Santa Cruz, and Anacapa from west to east—stretching west in a line out to sea from the end of the Santa Monica Mountains near Los Angeles. San Miguel faces six thousand miles of open Pacific Ocean, and thus gets blasted by some of the fiercest winds and largest waves anywhere in Southern California. The day I hiked across the island was typical, with a fierce wind yanking at my hat, shooing sand across the dunes, and tearing spray off the cresting swells. Winding my way down through the dunes to Simonton Cove, on the island's western shore, I found some very special pebbles in beach outcrops scrubbed clean by the waves. The pebbles were smooth and round, and on average about the size of a baseball. They lay encased in upended layers of Eocene* sand and gravel. They were not, I knew, native to San Miguel Island—or even to California. These pebbles were emigrants from Mexico.

Rock made of many smooth, rounded pebbles is called conglomerate, and it forms wherever breaking waves or flowing rivers tumble rock pieces and wear them smooth. Conglomerate is a common rock, but the pebbles in the conglomerate on San Miguel Island contain an uncommon curiosity: distinctive purple-maroon pebbles of rhyolite—a type of volcanic rock—sparkling with crystals of quartz and feldspar (figure 1.1). The rhyolite pebbles are so distinct—both visually and in their detailed chemical makeup—that geologists can confidently trace their origin to the exact volcanoes from which they eroded. Incredibly, those volcanoes, now long dead, are in Sonora, Mexico—*more than five hundred miles from San Miguel Island.* And that's not all. In the San Diego suburb of Poway, and in sea cliffs by La Jolla, you can find pebbles that are dead ringers for the ones on San Miguel Island. These, too, could only have come from near the same volcanoes in Sonora. How

* The Eocene period extends from fifty-six million to thirty-four million years ago. See the Geologic Time Scale at the front of the book.

FIGURE 1.1. (Upper) Eocene-age conglomerate from Mexico, now on San Miguel Island. We know that this rock traveled here from Mexico because it contains distinct purple-maroon volcanic pebbles that could only have eroded out of volcanic lava beds more than five hundred miles away in Sonora. The surrounding geology indicates that the pebbles were flushed into the ocean as part of a deep-water delta near the mouth of the Eocene (and now long extinct) Ballena River. (lower) Close-up with my finger pointing at one of the purple-maroon pebbles. The outcrop is part of the Cañada Formation in Simonton Cove on the northwest side of San Miguel Island. (Photographs by the author.)

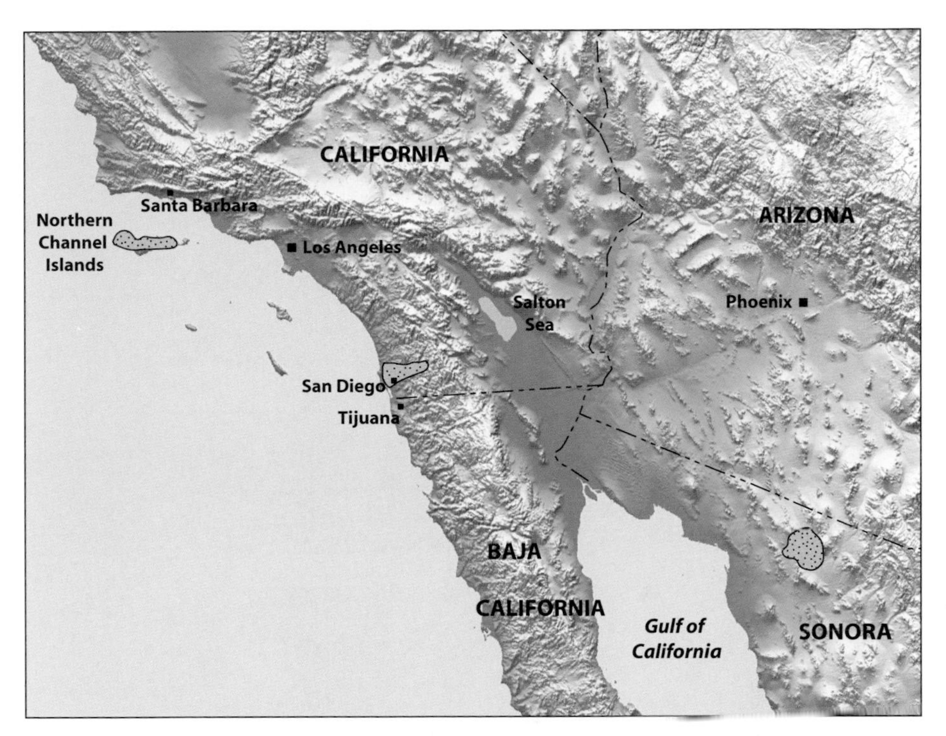

FIGURE 1.2. Distinctive Eocene-age riverbed pebbles on the Northern Channel Islands are also found around San Diego and near their source volcanoes in Sonora, Mexico (stippled yellow areas). The pebbles represent the remains of the forty-million-year-old Ballena River that was sliced apart by faults as the Pacific Plate slid northwest past the North American Plate. Figure 1.4 portrays the tectonic events that led to the pebbles' current distribution. (Shaded relief base from NASA, with labels added.)

did pebbles eroded from Mexican volcanoes migrate two hundred fifty miles to San Diego and five hundred miles to San Miguel Island (figure 1.2)?

The answer lies in the relentless creep of the Earth's tectonic plates. San Miguel Island today lies mostly on the Pacific Plate. The remains of the Mexican volcanoes lie on the North American Plate. San Diego occupies the fault-slivered zone in between (figure 1.3). The Pacific and North American Plates are sliding side-by-side past each other about two inches per year (a number first determined by matching up rock bodies split when the Gulf of California began to open about 5.5 million years ago, and confirmed today by global positioning system [GPS] measurements). That side-by-side shifting has carried the Mexican

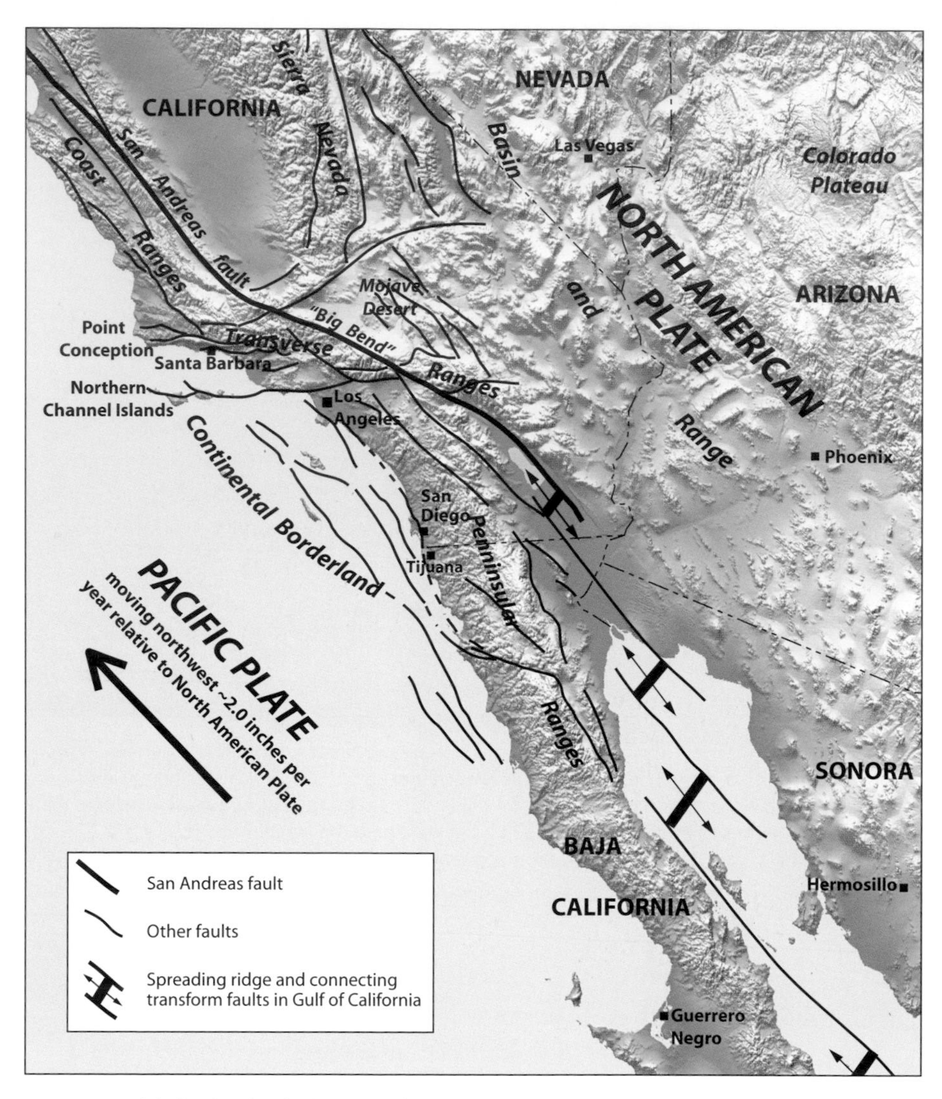

FIGURE 1.3. A belt of active faults more than two hundred miles wide marks the boundary between the Pacific and North American tectonic plates in Southern California. Movements along these faults have created most of the region's mountains, valleys, islands, and offshore basins. As the Pacific Plate heads northwest, it runs into the Big Bend in the San Andreas fault to create the Big Squeeze—a region of colossal compression that, among other things, is actively pushing up the Transverse Ranges. We'll explore the story of the San Andreas fault and the Big Bend/Big Squeeze in chapter 3. (Shaded relief base from NASA, with labels added.)

pebbles to where they are now. The story (summed up in figure 1.4) goes like this: About forty million years ago, a now extinct river, known to geologists as the Ballena River, flowed southwest from the Mexican volcanoes, carrying the distinctive pebbles toward the ocean. (The Gulf of California had not yet opened, so the river flowed uninterrupted to the Pacific.) About eighteen million years ago, sidling movements between the Pacific and North American plates began to split the pebbly deposits of the old riverbed. The easternmost section of the riverbed remained near its source in Sonora. The middle section slid northwest to where San Diego is now. The western section—the part of the river that poured into the ocean to form a deep-water delta—slid farther northwest, to where San Miguel and the other Northern Channel Islands are today (figure 1.4).

This story links directly to Southern California's large-scale geologic evolution, the details of which I'll give you in chapter 4. But you may be wondering: What's the connection here with deep time? Remember that the Pacific Plate slides past the North American Plate at just two inches per year—a spectacularly slow rate in human time. (A snail that fast would take three and a half centuries to cross my sixty-foot-wide suburban backyard.) But watch what happens when deep time comes into the picture. Two inches per year times eighteen million years (about how long ago the side-by-side movements between the two plates began in Southern California) equals 36 million inches, which is 540 miles—or almost exactly the distance that the pebbles on San Miguel Island now lie from their source volcanoes in Mexico. The message is as simple as it is powerful: Processes that seem trivially slow over human time can accomplish stunning work over geologic time.

THE PACIFIC PLATE–NORTH AMERICAN PLATE BOUNDARY

The story of the migrant pebbles on San Miguel Island highlights the most important geologic force at work in Southern California: side-by-side sliding of large blocks of the Earth's crust along big faults. Figure 1.3 shows that most of these faults trend northwest–southeast. This makes sense; that alignment allows the Pacific Plate to slide northwest past the North American Plate. Earthquakes happen whenever movements between the two plates cause one of these faults, every now and then, to snap. The San Andreas fault is the longest and most important—it's the big gorilla of California's faults—but the San Andreas doesn't act alone. Dozens of faults, spread across a zone more than two

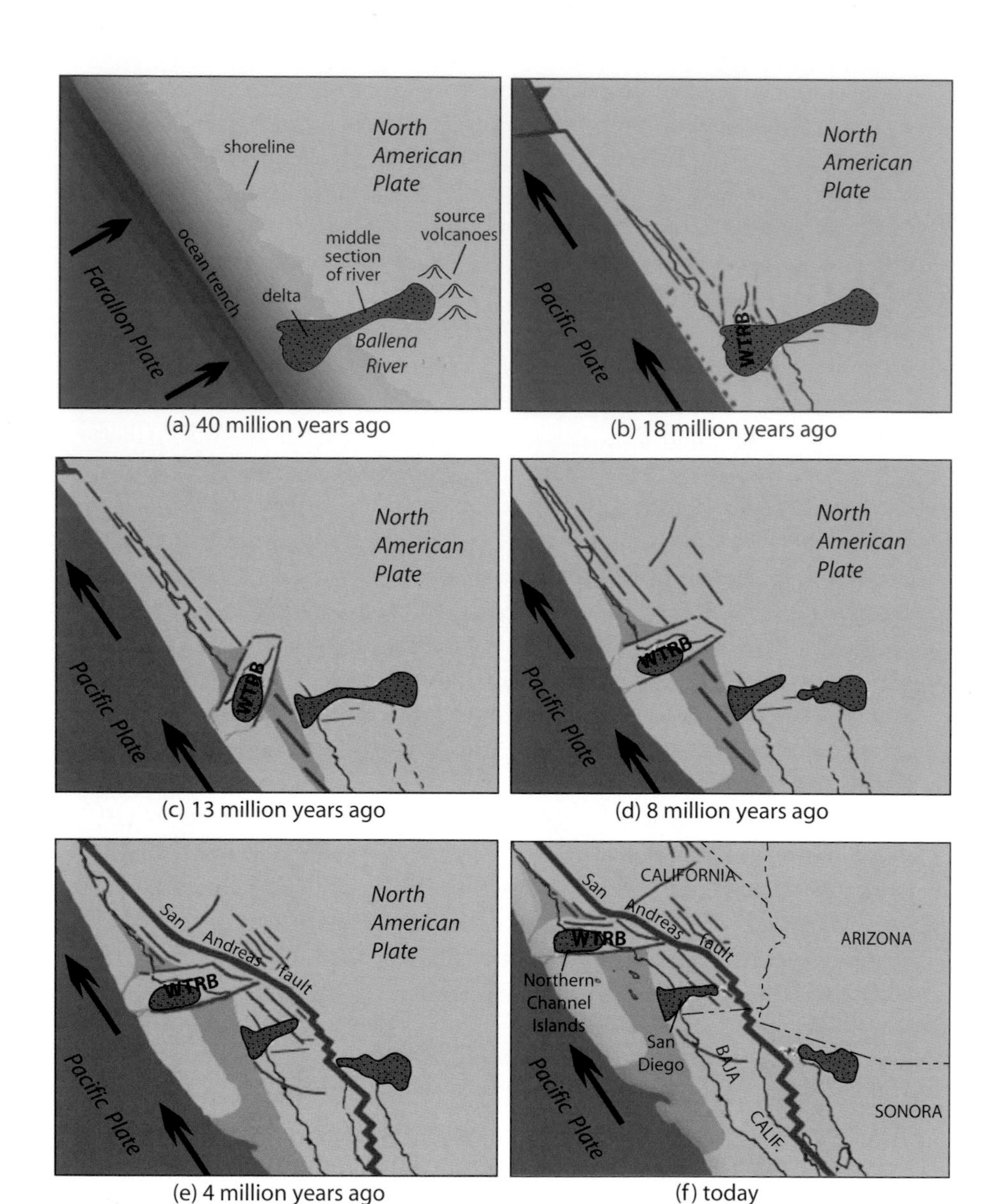

FIGURE 1.4. How faults and plate movements sliced up the remains of the ancient Ballena River and, in the process, created Southern California's present geography. Red lines mark faults. "WTRB" marks the Western Transverse Ranges Block of crust, which includes the Northern Channel Islands, the Santa Barbara Channel, and the Santa Monica and Santa Ynez mountains. The WTRB once lay near San Diego before the Pacific Plate captured it, spun it clockwise, and shipped it northwest. We'll look at the details and the supporting evidence for these developments in chapter 4. (Adapted from an animation by Tanya Atwater, University of California, Santa Barbara.)

hundred miles wide, allow the side-by-side movement of the two plates to happen.

This idea—that the boundary between the Pacific and North American plates is a wide zone of shifting faults rather than a single fault—is vital for understanding the geology not just of Southern California, but of the entire western United States. To see what I mean, look at figure 1.5, which shows the results of precision GPS measurements made at various places across the western United States over the past several decades. The arrows show how fast various places are moving in relation to the continental interior. One way to visualize this is to imagine pounding a huge nail through, say, Kansas to pin North America in place; those arrows in figure 1.5 show how various areas west of the Rockies would *still move.* (In other words, western North America is slowly tearing apart—a topic to which I'll return in a moment.) You can see that the fastest movements (longest arrows) lie on the Pacific Plate, showing that it moves northwest about two inches per year in relation to the continental interior. Slightly to the east, notice the cluster of arrows on the area marked as the Sierran Plate, showing that it moves northwest about one-half inch per year. The Sierran Plate,* which includes California's Great Central Valley and Sierra Nevada, seems to be dislodging from the rest of the continent because of what we call the *Big Bend* in the San Andreas fault (marked in figure 1.3). Because of the curve in the fault at the Big Bend, the Pacific Plate pushes like a shrugging shoulder against the south end of the Sierran Plate, shoving it north and cracking it away from the land to the east. But as the arrows in figure 1.5 show, the Pacific Plate is moving northwest faster than the Sierran Plate, and that means the rocks along the Big Bend are being severely crushed. That gives the Big Bend region of the San Andreas fault its other geologic moniker: the *Big Squeeze.* The crushing forces within the Big Squeeze make it one of California's most earthquake-prone regions—a story that we'll explore in chapter 3.

Besides earthquakes, another product of the Big Bend/Big Squeeze is the Transverse Ranges, shown in detail on the map of Southern California at the front of the book. The Transverse Ranges—which include the Santa

* You won't find the Sierran Plate marked in most geology textbooks, but like all plates, it is a large, coherent block of lithosphere (the outer rigid rock layer of the Earth that is broken up into moving plates) that moves on its own, its edges marked by zones of active earthquakes. The San Andreas fault system marks the west edge of the Sierran Plate. Its east edge is marked by another belt of active faults called the Walker Lane–Eastern California Shear Zone (figure 1.5).

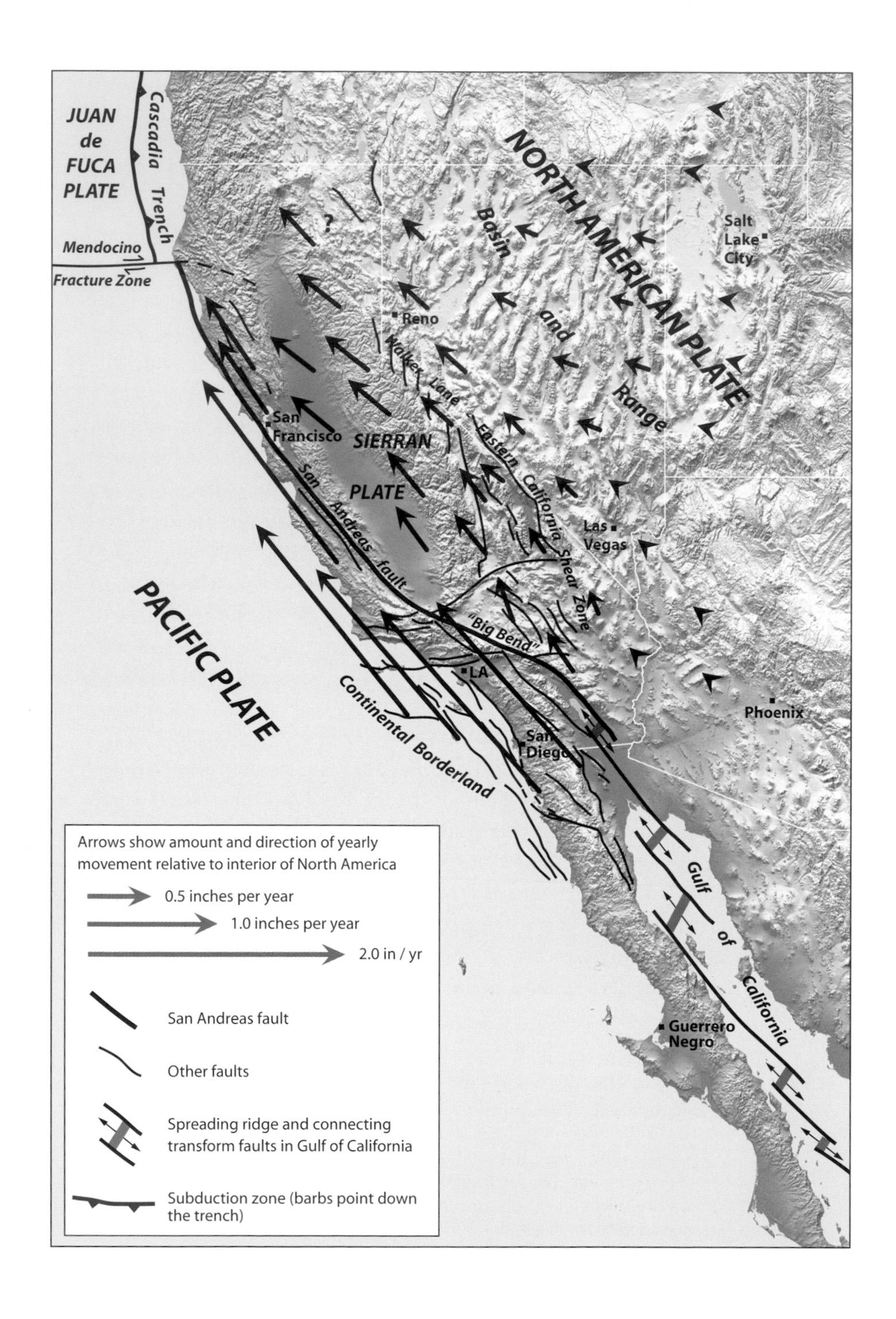

JUAN de FUCA PLATE
Cascadia Trench
Mendocino Fracture Zone
NORTH AMERICAN PLATE
Basin and Range
Salt Lake City
Reno
Walker Lane
Eastern California Shear Zone
San Francisco
SIERRAN PLATE
San Andreas fault
Las Vegas
"Big Bend"
LA
Phoenix
San Diego
PACIFIC PLATE
Continental Borderland
Gulf of California
Guerrero Negro
Arrows show amount and direction of yearly movement relative to interior of North America
0.5 inches per year
1.0 inches per year
2.0 in / yr
San Andreas fault
Other faults
Spreading ridge and connecting transform faults in Gulf of California
Subduction zone (barbs point down the trench)

Monica, Santa Ynez, San Gabriel, and San Bernardino mountains—get their name from their east–west alignment, transverse to the mostly northwest–southeast alignment of other mountains in California. Pinch a watermelon seed between your thumb and forefinger. Your thumb represents the Pacific Plate and your forefinger the Sierran Plate, pushing against each other in the Big Squeeze. The seed is the Transverse Ranges, popping upward to escape the pressure. Some parts of the Transverse Ranges are rising nearly one-half inch per year, making them the fastest-growing mountains in North America. They are also—not coincidentally—some of the steepest, most landslide-prone, and most earthquake-prone mountains in the nation.*

Returning to figure 1.5 and continuing east, notice the arrows across the Basin and Range Province, which spreads across all of Nevada and parts of neighboring states. The lengths of the arrows decrease eastward, telling us that points on the west side of the Basin and Range are moving northwest faster than points on the east side. In other words, the entire Basin and Range is *stretching*—the opposite of what is happening in the Big Squeeze. That stretching, which began some fifteen to twenty million years ago, explains why the Basin and Range looks the way that it does. The region gets its name from its washboard tempo of mountain ranges and intervening basins (valleys), all lined up generally north–south and divided from one another by large faults. Stretch the Earth's crust east–west, and it will break apart along north–south

* For the story of landslides from the San Gabriel Mountains laying waste to the suburbs of northern Los Angeles, see "Los Angeles against the Mountains" in John McPhee's book *The Control of Nature*.

FIGURE 1.5 (OPPOSITE). Western North America is slowly tearing apart across a belt of active faults that stretches from offshore California east to Utah and Arizona. The lengths of the arrows reflect the yearly rates of movement of particular points in relation to the continental interior, based on global positioning system (GPS) measurements reported by Kreemer et al. (2012). Regions west of the San Andreas fault, which include Baja California and the San Diego and Los Angeles areas, are mostly attached to the Pacific Plate and are moving northwest as much as two inches per year in relation to the interior of North America. The Sierran Plate, which lies between the San Andreas and Walker Lane fault systems, moves northwest a little more than a half-inch per year. The Basin and Range has already stretched by more than two hundred miles since its inception some fifteen to twenty million years ago, and it continues to widen today by nearly a half-inch per year. As the western United States breaks apart, one or more ocean basins may eventually open up through California or Nevada, as shown in figure 1.6. (Shaded relief base from NASA, with labels added.)

aligned faults. Blocks of rock that are rising along these faults make the ranges of the Basin and Range, while blocks that are dropping make the basins (valleys) in between. Since its inception, the Basin and Range has stretched by *more than two hundred miles* in some areas. And the work goes on. Each year, the drive between Reno and Salt Lake City increases by about half an inch (GPS measurements prove it). The American West is a living landscape, reshaping itself a bit every year as the Pacific Plate grinds northwest past the North American Plate and drags pieces of our continent along with it.

Before we return to Southern California, let me speculate a bit on what those movements shown in figure 1.5 might mean for the future of the western United States. In figure 1.6, I propose three not-too-fanciful scenarios for what the western part of the country may look like in the geologic future. Each scenario portrays a possible geography about fifteen million years from now. Each assumes that the Pacific Plate will continue to scrape northwest past the North American Plate at its present rate of two inches per year.

In scenario 1 in figure 1.6, the San Andreas fault takes over as the main locus of side-by-side motion between the two plates. Baja California and coastal California, including the Los Angeles and San Diego areas, shear away from the rest of the continent to form a long, skinny island. A short ferry ride across the San Andreas Strait connects Los Angeles to San Francisco.

In scenario 2, the San Andreas fault sputters out, and the Walker Lane–Eastern California Shear Zone (marked in figure 1.5) takes over as the main locus of motion between the two plates. All of California west of the Sierra Nevada, together with Baja California, shears away from the rest of the continent. The Gulf of California extends north like a growing wedge to become the Reno Sea, which divides California from western Nevada. Residents of Nevada gambol and gamble along the shores of their new-formed ocean. The scene is reminiscent of how the Arabian Peninsula split from Africa to open the Red Sea some five million years ago.

In scenario 3, central Nevada splits open through the middle of the Basin and Range, where the highly stretched crust is already thin and weak. The widening Gulf of Nevada divides the continent from a large peninsula composed of Washington, Oregon, California, Baja California, and western Nevada. The scene is somewhat like Madagascar's origin when it split from eastern Africa to open the Mozambique Channel.

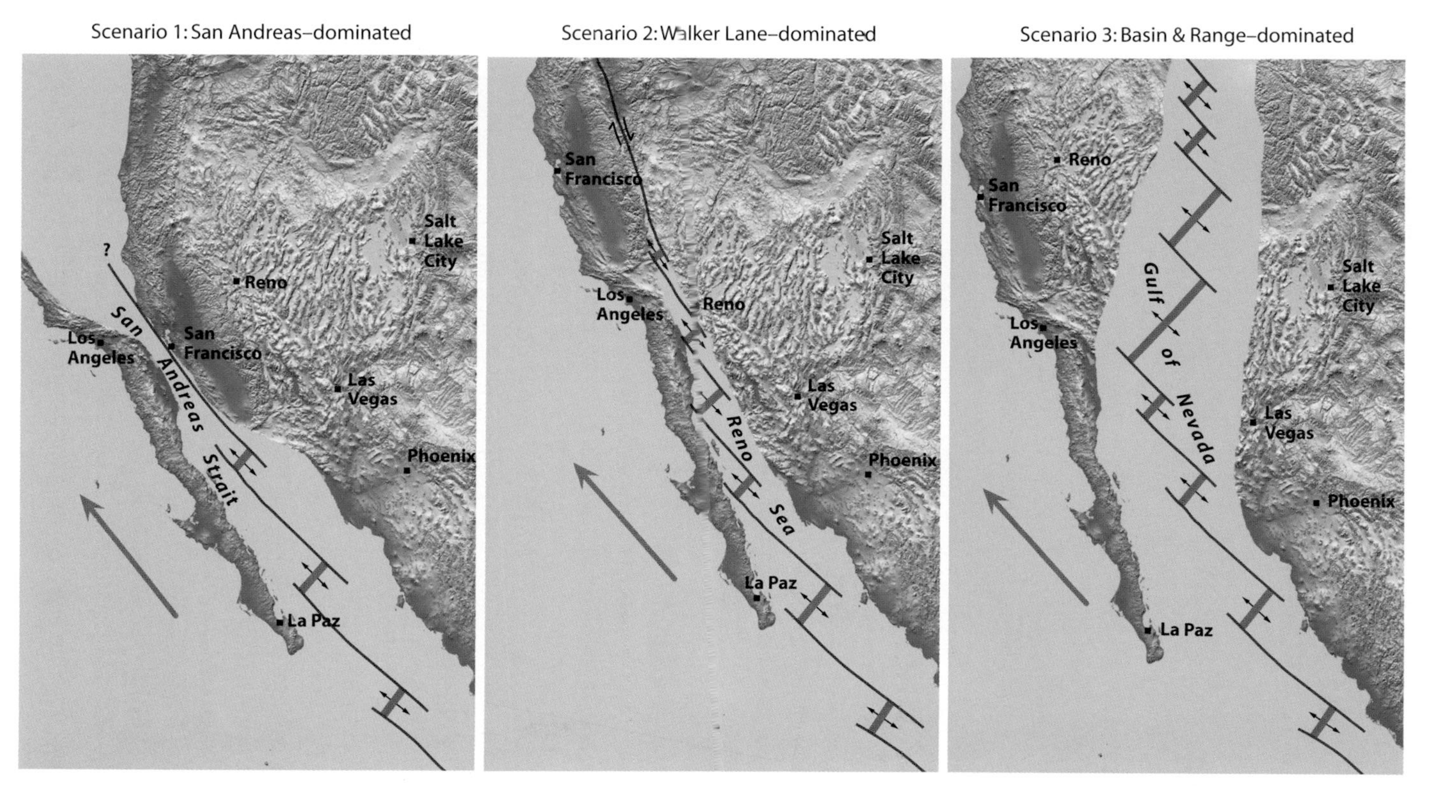

FIGURE 1.6. Possible geographies of the American West some fifteen million years in the future. (Shaded relief base from NASA, with labels added; from Meldahl 2011.)

Those scenarios encompass a range of possible outcomes; the actual result may combine all three. But every projection points to one conclusion: continental fragmentation, and eventual beachfront property in the deserts of the American West.*

EARTHQUAKES WITHIN THE PLATE BOUNDARY

Returning our focus now to Southern California, let's consider what that two-hundred-mile-wide, fault-fractured boundary between the Pacific and North American plates means for us today. One consequence is earthquakes.

Southern California is a geologic work-in-progress, and earthquakes are its growing pains. More than any other force, earthquakes have created Southern California's landscape. This may seem odd, given that most of us—even if we live here for years—will probably feel only a few moderate shakes and perhaps never experience a "big one." But that speaks mostly to the limits of human time. If we stretch our view out across geologic time, our perspective changes. Over millions of years, the cumulative work of numerous earthquakes can do stupendous landscaping. Practically all the mountains, valleys, islands, and deep offshore basins of Southern California exist *because* of earthquakes—many millions of earthquakes— each doing its part to shift a portion of the Earth's crust up, down, or sideways a few inches (small quake) or a few feet (large quake) at a time. Think back to those pebbles on San Miguel Island, and imagine the number of earthquakes it must have taken to move those rocks, lurch by lurch, five hundred miles from Mexico.

Southern California experiences thousands of earthquakes every year. Most are too small for any of us to feel, but seismometers (sensitive ground-motion instruments) may pick up a dozen or more quakes *every day* throughout Southern California. Large quakes are much rarer than small ones. A very large quake may not happen in Southern California for fifty years, or one may rip loose in the next five seconds—we can't know. (My money says it'll be sooner rather than later, and I'll tell you why in chapter 3.) The only certainty is that earthquakes, large and small, have been crackling across Southern California for millions of years and will continue. Earthquakes are our constant reminder that much of California (especially the area west of the San Andreas fault)

* These ideas are adopted from my book *Rough-Hewn Land: A Geologic Journey from California to the Rocky Mountains* (University of California Press).

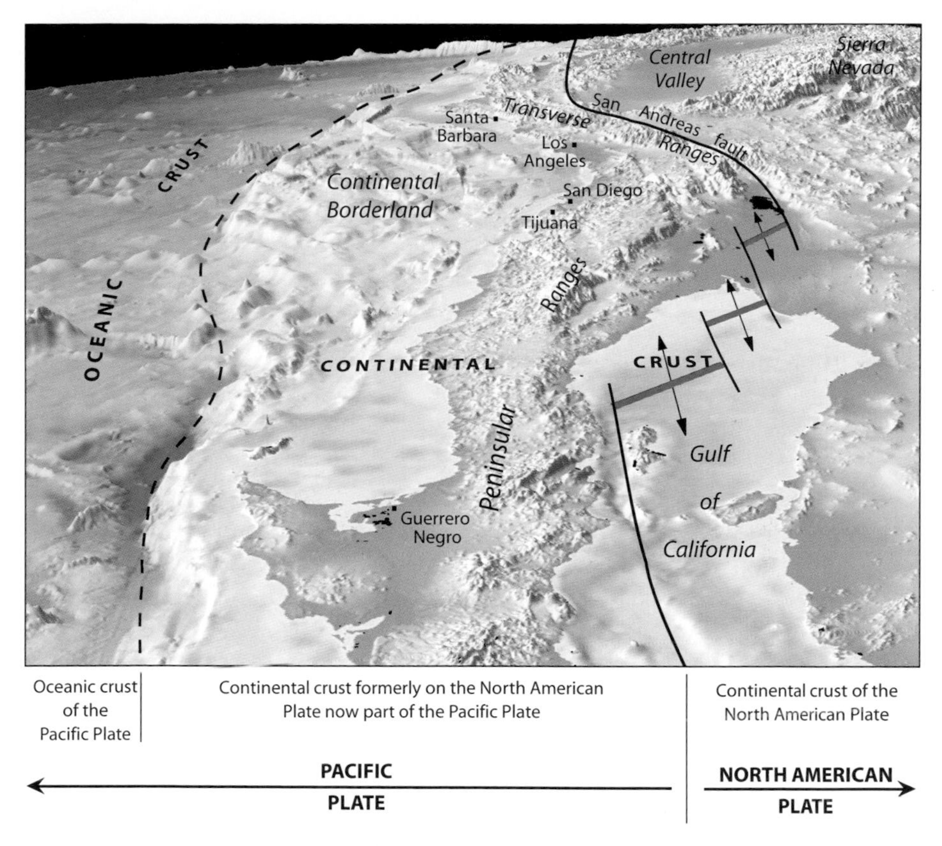

FIGURE 1.7. This image looks obliquely north-northwest from above Baja California toward Southern California, with the ocean removed and colors representing elevations and water depths. The dashed line to the west (left) divides continental crust (which underlies land and shallow ocean areas) from oceanic crust (which underlies the deep Pacific). Notice that the boundary between continental and oceanic crust is not the same as the boundary between the Pacific and North American tectonic plates, which passes north up the Gulf of California to merge with the San Andreas fault. This is because the Pacific Plate contains large portions of continental crust that were *once part of North America.* These portions include Baja California, the Continental Borderland, and coastal California west of the San Andreas fault. In chapter 4, we'll explore how these former pieces of North America ended up hitching a ride with the Pacific Plate. (Color relief base from GeoMapApp, www.geomapapp.org, with labels added.)

doesn't belong to North America. We ride mostly with the Pacific Plate, and earthquakes are a direct consequence of the Pacific Plate hauling big portions of western California toward Alaska (see figure 1.7). The long-term fate of coastal Southern California, tectonically speaking, is linked more to Hawaii, Easter Island, and other places on the Pacific Plate than to, say, El Centro or Bakersfield.

The Sunken Continental Borderland

Will coastal California sink into the ocean if it splits away from the rest of the continent along the San Andreas fault? I hear this question surprisingly often. The notion may trace back to the 1978 movie *Superman,* starring Christopher Reeve as Superman and Gene Hackman as Lex Luthor, the villain. Lex Luthor's get-rich scheme is to buy large tracts of cheap land in the California desert east of the San Andreas fault. He then plans to explode a nuclear missile along the fault, which will dump western California into the sea and make him the owner of hundreds of miles of sparkling new beachfront property. (Spoiler alert: Superman stops him.)

Luthor's plan had a fatal flaw. Continents are made of relatively thick, low-density rock that floats buoyantly in the denser rock of the mantle beneath (see figure). A branch that breaks away from a floating log won't sink; its buoyancy doesn't depend on being attached to the log. Likewise, a piece of continent that splits away from the rest won't sink into the mantle; its buoyancy doesn't depend on being attached to the rest of the continent. The buoyancy of thick, light continental rock keeps the continents above the sea in most places. By contrast, the denser, thinner rock that forms the deep ocean floor floats low in the mantle and thus lies well below sea level. (The concept that the thickness and density of crustal rock controls how high or low it floats in the mantle, and thus its elevation, is called *isostatic equilibrium.*)

Having said that, there *is* a way to make continental rock founder at least partly below sea level. It can happen where faults stretch and thin the continental crust. Just as a thick iceberg will rise higher above the water than a thin one, so a thick piece of continental crust will float higher in the mantle than a thin piece. If a thick iceberg breaks apart into smaller, thinner pieces, none of those will float as high as the original. Likewise, if a thick piece of continental crust stretches and breaks into thinner pieces, none of those crustal pieces will float as high in the mantle as before. This is why the Continental Borderland is mostly below sea level—despite being made largely of conti-

THE SOUTHERN CALIFORNIA BIGHT AND THE CONTINENTAL BORDERLAND

Now that we've explored some aspects of Southern California's plates, faults, and earthquakes, let me introduce some of the region's coastal geology and oceanographic features (as a prelude to deeper detail to come in later chapters).

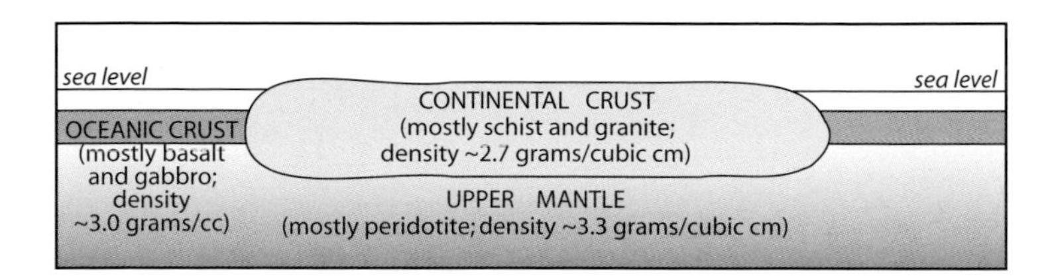

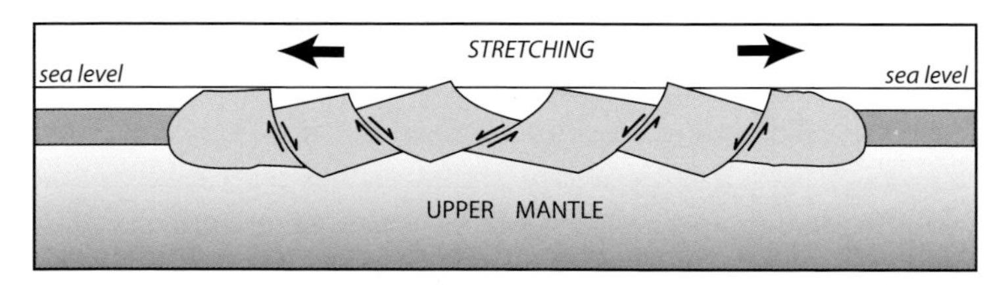

The Earth's continental crust is composed of rock that is thicker and less dense than oceanic crust—the rock of the deep ocean floor. This makes the continents float high in the Earth's mantle, so that the continents rise above the ocean surface in most parts of the world (upper diagram). But if the continental crust stretches and breaks up along faults, parts of it may sink below sea level, leaving just a few areas standing high enough to form islands. That is what we see in the Continental Borderland (lower diagram).

nental crust. As the Pacific Plate, sliding northwest, tore away pieces of the North American continent to create the Continental Borderland (a story we'll explore in chapter 4), the crust stretched and broke apart into several blocks (lower image on the figure). Today, the highest parts of some of these blocks poke above the sea as islands or lie barely awash as shoals like Cortes Bank, Tanner Bank, Thirty-Mile Bank, and others. The deepest parts of the foundered blocks create basins like the Santa Barbara, San Nicolas, Santa Monica, and Catalina basins. For a fuller view of the broken-up continental crust of the Continental Borderland, see figure 1.7.

From Oregon south to Point Conception, the California coast faces generally west. But at Point Conception the coast takes a right-angle turn to the east, so that someone standing on the beach at Santa Barbara looks south toward the ocean, not west. From Point Conception to San Diego, the shape of the coast describes a broad curve, 260 miles long, looking somewhat like a bite out of a big cookie. This is the

Southern California Bight, shown in detail on the map of Southern California at the front of the book. ("Bight" is a nautical term for a large inward curve in a coast that is larger and less confined than a bay.)

The Southern California Bight faces a collection of islands, shallow banks, and deep basins offshore called the Continental Borderland (see figure 1.7 and the map of Southern California). As the name suggests, the rocks of the Continental Borderland, although largely underwater today, are mostly pieces of continental crust (see text box). In fact, the Continental Borderland was once part of the North American Plate, as were Baja California and much of California west of the San Andreas fault. As I explained above, these former pieces of North America today are traveling mostly with the Pacific Plate. Why did they shift over? The answer is that the Pacific Plate kidnapped them (a story we'll explore in chapter 4). As the Pacific Plate dragged the rocks of the Continental Borderland northwest, it stretched them and broke them up into multiple blocks. The shifting of these blocks—up, down, and sideways along faults—has created the Continental Borderland's rugged topography (mostly underwater) of islands, shallow banks, undersea ridges, and deep basins (see text box for more detail). If you could drain the ocean and fly across the Continental Borderland, you would see a fault-fractured landscape not unlike Nevada's Basin and Range. That's because both regions formed in much the same way—by wide-scale stretching of the crust. Flood Nevada so that only the highest peaks poke out as islands, ring those islands with forests of kelp, and you have a fair image of the Continental Borderland. The Patton Escarpment (see the map of Southern California at the front of the book) marks the western edge of the Borderland. That's where the seafloor geology shifts from fault-shattered blocks of continental crust to the oceanic crust of the Pacific Plate. Figure 1.7 portrays this shift clearly.

CURRENTS AND WAVES IN THE SOUTHERN CALIFORNIA BIGHT

The Southern California Bight and the Continental Borderland make oceanographic conditions in Southern California different from those in areas to the north. The water in the bight is warmer, and wave behavior more complex, than north of Point Conception. The reasons come down to ocean currents and the blocking of ocean swells by the islands. The story begins with the California Current.

The California Current flows south along North America's west coast about ten miles per day, with a volume greater than fifty Amazon Rivers. The current carries cold water south as part of a great loop of cur-

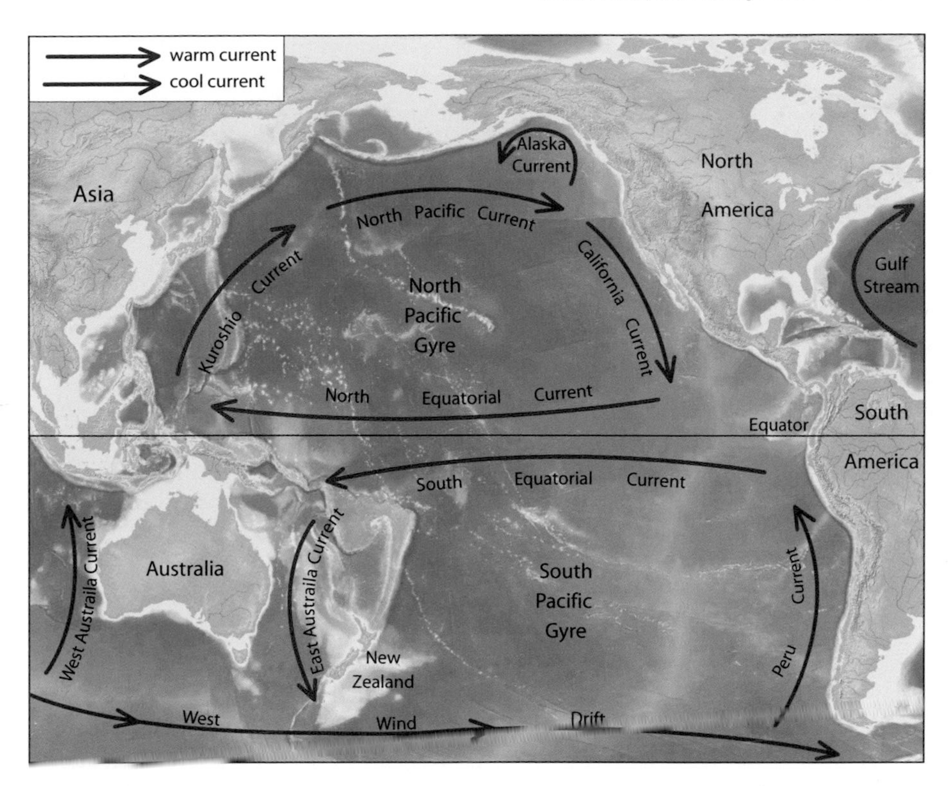

FIGURE 1.8. The world's ocean currents form large loops, called *gyres,* driven by major belts of prevailing winds. The California Current forms part of the North Pacific Gyre. The current carries cold water from the North Pacific toward the equator. For closer views of the California Current, see figures 1.9 and 1.10.

rents called the North Pacific Gyre, which cycles clockwise around the North Pacific. A water molecule takes about five to six years, on average, to travel the whole way around the gyre (figure 1.8). Along most of California's coast, the California Current flows close to shore, keeping coastal waters cold and sending cool air and abundant fog inland. But where the coast angles east at Point Conception, the current continues south, taking it far from the mainland. By the latitude of San Diego, the current is more than 150 miles offshore. I sometimes hear kayakers and surfers along San Diego's beaches talking about how the "California Current" carried them south along the beach, but this local shore-parallel current—called a *longshore current*—has no connection to the much larger California Current 150 miles to the west. (The longshore current flows in whatever direction the waves happen to be going as they

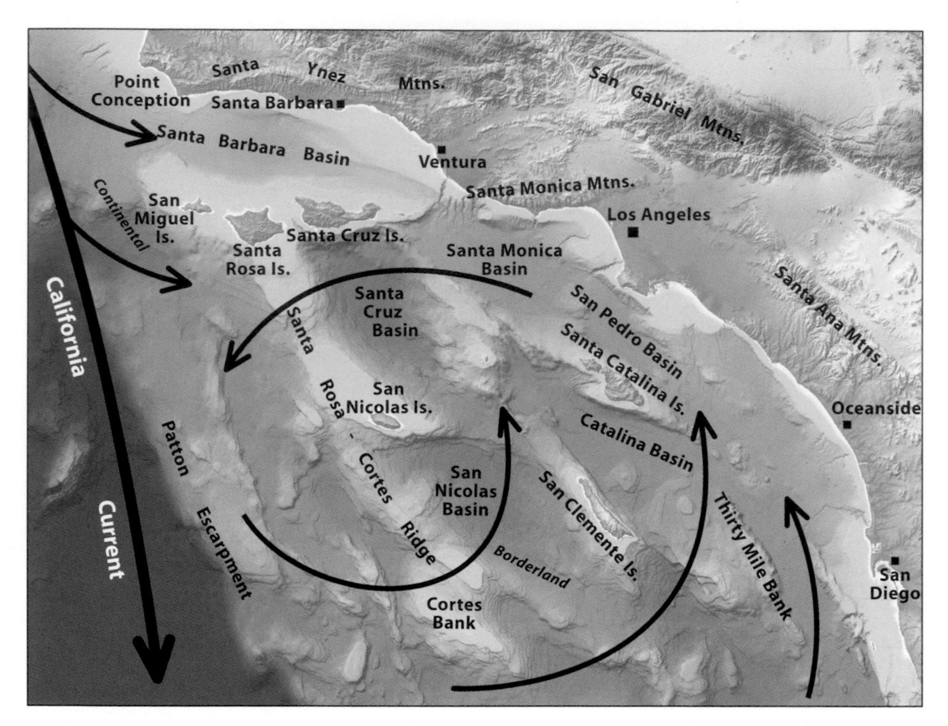

FIGURE 1.9. Ocean circulation within the Southern California Bight. The eddy pattern shown is highly generalized; actual circulation is complex, with many small eddies forming and unraveling within the larger cycle. Some water from the California Current enters the Southern California Bight at its north end, but mostly the California Current stays west of the Continental Borderland, the western edge of which is marked by the Patton Escarpment. The shore-parallel currents that you may feel along a local beach are not related to the currents here. Those are longshore currents, and they flow along the beach in whatever direction the waves happen to be heading that day. Longshore currents do not extend seaward past the surf zone (the zone where the waves are breaking). (Shaded relief base from NOAA, with labels added; based on Hickey 1992.)

approach the beach at an angle, and that can change from day to day.) As the California Current flows along the edge of the Patton Escarpment, it drags past the water in the Southern California Bight, setting up large counterclockwise eddies (figure 1.9). This eddy circulation means that new water comes into the bight from both the north (cool water) and the south (warm water). This, along with the eddies corralling water within the bight, makes bight waters consistently warmer than anywhere else along the California coast, as you can see in figure 1.10.

Ocean waves approach the Southern California Bight from all directions in the Pacific, but most commonly from the west and northwest.

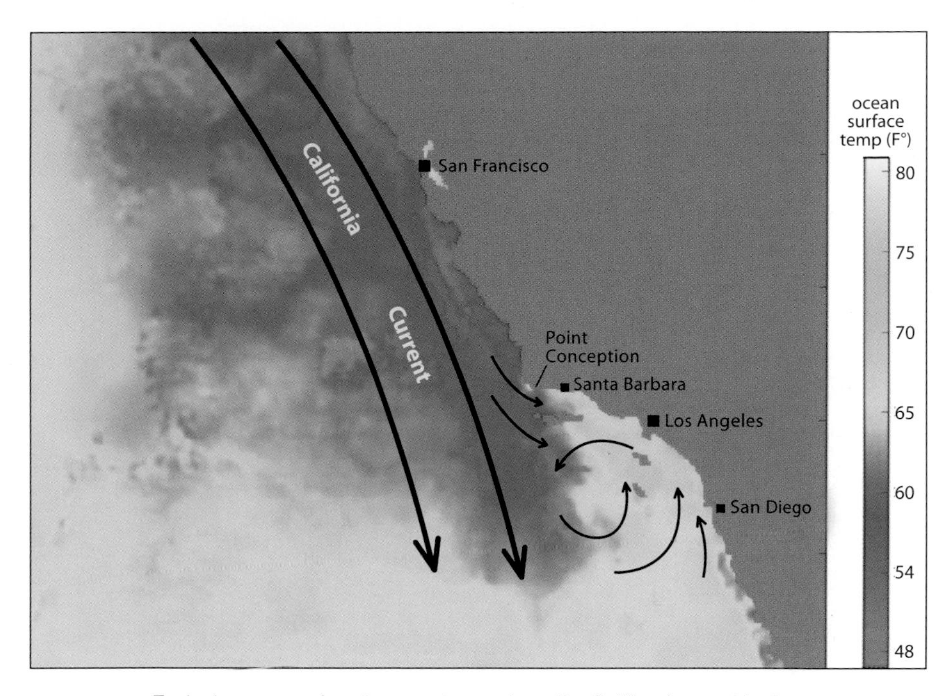

FIGURE 1.10. Typical ocean surface temperatures along the California coast in August (temperature scale on right). North of Point Conception, the cold California Current flows close to land, keeping coastal waters cool year round. Water temperatures drop even more when deep waters upwell along the coast (purple areas), which happens whenever surface waters are pushed away from land by prevailing winds and the Earth's rotation (Coriolis effect). Where the coast angles east at Point Conception, the California Current continues south, taking it far from the mainland. Waters in the Southern California Bight are therefore consistently warmer than waters north of Point Conception. (Image from NASA's Jet Propulsion Laboratory, http://ourocean.jpl.nasa.gov/SST, with labels added.)

That's because storms frequently whip up large waves in the northernmost Pacific, particularly in the late fall, winter, and early spring. As these northwest swells approach the bight, two things control their power and behavior. First is the orientation of the coastline. The southwest-facing shoreline of the bight reduces the impact of west and especially of northwest approaching swells, which are partially blocked by Point Conception. By contrast, south and southwest swells (most common in summer) hit the bight more-or-less straight on. Second, the bight faces the maze of islands and shallow banks that make up the Continental Borderland. Just as pieces of furniture in a room cast shadows from a sunbeam, so the islands cast *swell shadows* (areas of smaller waves behind the islands) across some parts of the bight, while opening *swell*

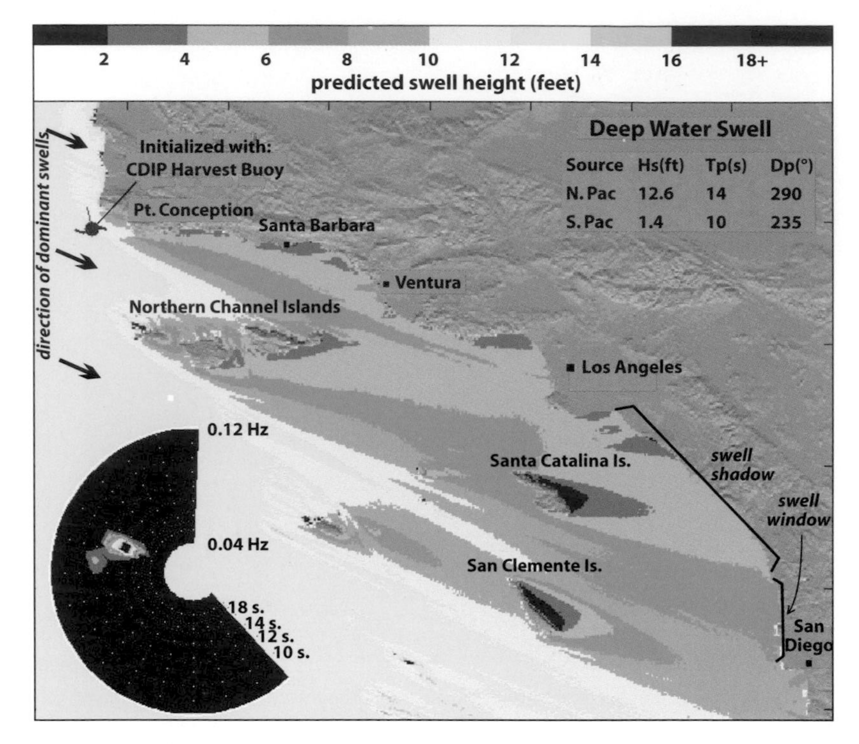

FIGURE 1.11. The direction of wave approach and the shadowing effects of islands control wave sizes in the Southern California Bight. The color bar corresponds to predicted wave heights. On the day shown, deep-water swells with a period (i.e., time between successive wave crests) of 14 seconds and a height of 12.6 feet are approaching from compass direction 290 degrees (20 degrees north of due west). The largest waves are along the coast north of Point Conception because the coast there faces the oncoming swells. Coastal wave heights drop dramatically south of Point Conception, both because the coastline faces southwest (and thus receives the northwest swells indirectly) and because the islands and shallow banks of the Continental Borderland dampen approaching waves. Notice the *swell shadow* on the mainland cast by the sheltering effect of Santa Catalina Island and the Northern Channel Islands, and the *swell window* near San Diego formed by waves approaching through the gap between Santa Catalina and San Clemente islands. Swell shadows and swell windows control the location and quality of surfing waves in Southern California—a topic we'll explore in chapter 5. (Image from the Coastal Data Information Program, Scripps Institution of Oceanography, http://cdip.ucsd.edu/; with black text, lines, and arrows added.)

windows (areas where waves pass unimpeded between the islands) across other parts, as shown in figure 1.11. And just as the shifting angle of the Sun throughout the day causes parts of a room to go from light into shadow, and vice versa, so wave sizes in the bight change—sometimes daily—with shifts in the direction of arriving swells. This makes Southern California one of the most complex wave- and surf-forecasting regions on the west coast—a topic to which we'll return in chapter 5.

2

Tsunamis

I thought I was running for my life with the end of the world chasing me.

—A survivor of the March 2011 tsunami in Japan, as told to *The Washington Post*

The seaports of Los Angeles and Long Beach lie side-by-side at the head of San Pedro Bay, on the south side of greater Los Angeles. They are, respectively, the first- and second-busiest seaports in the United States, and together form the sixth busiest seaport complex in the world. Your shoes and shirts probably spent time there, along with numberless other goods in your daily life that were made in Asia—cars, computers, televisions, socks, chairs, chopsticks, lipsticks, fuel injectors, toys, pajamas, and so on. Billions of dollars in goods pass through the seaport complex each year, mostly arriving from, or going to, China, Japan, South Korea, Hong Kong, Thailand, and Taiwan. After tugboats shove the ponderous freighters into their berths, outsized cranes offload the shipping containers: steel boxes measuring eight feet high, eight feet wide, and twenty to forty-five feet long. Take all the shipping containers that move through the Los Angeles–Long Beach seaport complex each year and put them end-to-end, and they would stretch around the Earth almost two times. Shut the two seaports, and the American economy would lose nearly one billion dollars per day.

At the city of Miyako on Japan's northeast coast, a concrete wall thirty feet above sea level and more than a mile long faces the ocean. It was built to protect the town from tsunamis—waves spawned by sudden displacements of ocean water, such as when an undersea fault shifts during an earthquake. Miyako's residents, like those of other coastal Japanese towns, regularly practice tsunami drills. When the sirens go off, everyone scurries for the safe side of the wall. Then great steel gates, normally left open for traffic to and from the harbor, clang shut to seal off most of the town.

One hundred and twenty miles east of Miyako, the Pacific tectonic plate plunges into the Japan Trench at a speed of nearly four inches per year. For several centuries before March 11, 2011, the descending plate had been slowly pulling down the seafloor west of the trench. (Visualize reaching up from a swimming pool to pull down the end of a diving board.) At 2:46 P.M. local time on March 11, the Pacific Plate released its grip (visualize letting go of the diving board), and in a few seconds, a hundred-mile-long by thirty-mile-wide swath of seafloor snapped upward as much as twenty-five feet. As earthquake waves whipsawed Japan, tsunami sirens began wailing in dozens of coastal towns. Tsunami waves began radiating outward from the offshore disturbance at jetliner speeds. The first waves arrived at the nearest beaches within fifteen minutes of the quake, and over the next few hours, one wave after another washed ashore along several hundred miles of Japan's northeast coast. (Multiple surges and withdrawals are the rule in tsunamis, just as a rock tossed into a puddle causes several waves, not just one, to wash up against the shore.)

Miyako's residents had time to get ready. People cleared out of the harbor and other low-lying areas and collected behind the tsunami wall. As local police closed the gates, some residents gathered on rooftops with handheld video cameras. ("Hopefully we'll see something," a few must have thought. False alarms are the rule in most tsunami warnings, because most earthquakes don't make tsunamis.) The videos they shot would show the world what happened next. A rising tide of seawater, black with mud vacuumed off the seabed, surged into the harbor at thirty miles per hour, snapping boats from their moorings and sweeping cars off the docks. The sea rose so high, so fast, that when it met the thirty-foot-high tsunami wall it simply *poured over,* like a filthy Niagara Falls, laying waste to the town. Measurements later showed that the tsunami at Miyako had fifty-three feet of run-up—meaning that it reached fifty-three vertical feet above sea level. Run-ups of thirty feet or more inundated 250 miles of Japan's east coast that day—a distance equal to the coastline from San Diego to Point Conception. Parts of Iwate Prefecture, the hardest hit, experienced run-ups of more than a hundred feet. In all, the March 2011 earthquake and tsunami together killed 15,867 people, and another 2,909 are still missing and presumed dead.

FIGURE 2.1. Aerial view, looking southwest, of part of the Los Angeles and Long Beach seaports, with Santa Catalina Island in the distance. The breakwaters, designed to block ocean swells, would offer little protection from tsunamis triggered by earthquakes or submarine landslides in the Continental Borderland. (Photograph by Bruce Perry, California State University, Long Beach.)

An eight-mile-long set of breakwaters protects the Los Angeles and Long Beach seaports from ocean swells (figure 2.1). The breakwaters stand just fifteen feet above the sea.

A large tsunami will wash ashore in Southern California one day. For reasons I'll explain shortly, Southern California earthquakes aren't likely to spawn a tsunami nearly as large as the 2011 wave that hit northeastern Japan. But a tsunami with twenty feet of run-up is well within geologic expectation. Southern California's coast is similar to northeast Japan's in population density and topography; both regions have steep, rugged shorelines that alternate with low, densely populated coastal plains. Twenty feet of run-up would translate into more than a mile of inland penetration in some areas of Orange County, western Los Angeles, and the Oxnard Plain. Let me spin a bit of plausible fiction about how a tsunami disaster might unfold in Southern California, and then back up this fable with some science.

A TSUNAMI COMES TO SOUTHERN CALIFORNIA

Santa Catalina Island rises sharply from the sea twenty-five miles southeast of the seaports of Los Angeles and Long Beach. The island exists because of the Catalina fault, which, over the past several million years, has hoisted it, earthquake by earthquake, so that it now rises more than a vertical mile above the floor of the Catalina Basin. Virtually all the topography of the Continental Borderland—every island, bank, undersea ridge, and deep basin—is a product of shifting faults and earthquake violence, and every earthquake that rattles the region today testifies to its ongoing geologic reformation. Although practically invisible in human time, that reformation is as continuous and relentless as its root cause—the Pacific Plate's northwestward migration past North America (see figure 1.3).

We can credibly imagine that centuries of building stress have pushed the Catalina fault to its next breaking point. Without warning, the fault snaps, unleashing a magnitude 6.8 earthquake (see text box "Measuring the Strength of Earthquakes" in chapter 3 for an explanation of earthquake magnitudes). Catalina Island leaps three feet as the seabed to the west lurches several feet toward Alaska. Tourists strolling around Avalon, the island's largest community, are thrown to the ground, while buildings collapse throughout town. The seismic waves begin to thrash greater Los Angeles twenty seconds later. Angelenos duck and cover while holding their collective breath. The shaking lasts several minutes, but the damage on the mainland, which lies twenty-five or more miles from the epicenter, is not excessive: Some walls crack, some crockery smashes, and some bookshelves fall over. It isn't the Big One, Angelenos soon realize; in fact, it isn't even the San Andreas fault. (I'll come back to those topics—the Big One and the San Andreas fault—in chapter 3.) Life in LA returns to normal . . . for about ten minutes.

A few miles south of the Palos Verdes Peninsula, the continental shelf abruptly ends as the seafloor plunges more than a half-mile down the San Pedro Escarpment. For centuries, sand and mud have piled up on this precarious slope, and it now sits at hair-trigger instability. Shaken by the seismic waves, a titanic landslide—one-fifth of a cubic mile of sand and mud—sloughs down into the San Pedro Basin. A bulge of displaced water, several feet high and more than five miles across, wells up like a blister on the sea surface and begins to radiate outward. Over the deep San Pedro Basin, the tsunami waves are no danger. (In deep water, tsunamis are very fast but low and inconspicuous, and they often pass beneath ships unnoticed.) But where tsunamis approach land, the shallowing seabed slows them down and forces their heights upward, transforming them into freight trains of destruction. Moreover, as the waves head toward Long Beach, the curving continental shelf bends them like light beams through a magnifying glass, so that their full power is aimed at the seaport complex.

The leading wave of the tsunami surges over the breakwaters outside the seaports. The shallowing continental shelf has now slowed the wave to perhaps forty miles per hour, but in doing so has pushed up its height to twenty feet. It looks nothing like a normal ocean wave. Sailors on berthed freighters, and tourists strolling on the decks of the *Queen Mary,* get the first clear views. They see a thick, roiling tide, approaching fast, with a sloping face twenty feet high. The similarity to fast-rising tides explains the widespread misnomer "tidal wave," although tsunamis and true tides are unrelated. (The daily rise and fall of the sea that we call *tides* results from the Earth rotating under broad bulges of ocean shaped by the gravity of the Moon and Sun.)

The tsunami surges onto the seaport docks and the parking lots around the *Queen Mary,* sweeping up hundreds of cars in a bobbing flotilla. The rising flood lifts the ship, and she lists and tightens against her cables. With gunshot snaps, the cables part, and the ship heaves and rolls, tossing tourists to her decks. Lifted nearly twenty feet by the rising waters, she bashes through the rock barricades around her and sets forth on her first voyage since 1967, when she arrived in Long Beach Harbor to become an immobile tourist trap. Impelled by the tsunami, she speeds gracefully north up Queensway Bay (the outlet of the Los Angeles River) before she T-bones the Queensway Bridge. The steel bridge folds like a bent coat hanger around her bow. Cars and trucks pitch off into the water, falling through snowflakes of green paint that pop away from the bending steel.

More than a century ago, before the seaports of Los Angeles and Long Beach existed, the small fishing village of Wilmington went about its business on the shores of Wilmington Lagoon, at the head of San Pedro Bay. A long sand spit called Rattlesnake Island lay between the lagoon and the bay, forming "a colorful, smelly home for fishermen, artists, writers, some ne'er-do-wells, and the ever-present noisy, wheeling fish gulls." That sand spit—now vastly expanded with concrete and landfill—is Terminal Island, the heart of the LA–Long Beach seaport complex. The lagoon behind it is now dredged to freighter depths and lined with shipping cranes. The point is that the entire seaport complex began on land at sea level and remains so today. Dozens of tsunami evacuation signs (put up throughout coastal California after the great 2004 Indian Ocean tsunami) point to escape routes out of the seaport. But the traffic-clogged seaport is not a place you can leave quickly, even under normal circumstances. With a half-hour warning, you might be able to get out to high ground (or not, since panic tends to logjam escape routes). Depending on where a tsunami forms in the Continental Borderland, you could have as little as eight minutes.

The wave sweeps through the seaport complex, gathering up flotillas of shipping containers. Until they flood, closed containers are buoyant. The immense steel boxes tumble at the front of the wave like hundreds of battering rams, smashing everything in their path. Docked freighters pitch and yaw against their bollards, and some break free. The wave isn't quite high enough to carry them over the docks, so they bump upstream along the shipping channels, mashing smaller boats flat against the docksides. The shriek of bending steel and the boom of heavy objects tumbling join the rumble of the wave, creating a numbing cacophony. By the time the wave passes under the bridges leading to Terminal Island, it has gathered up so much debris that it looks, from afar, like chunky soup. Ship fuel spills from damaged storage tanks, and fires, triggered by ruptured gas lines, begin to break out across fuel-soaked patches of moving ocean.

Hundreds of dockworkers have swarmed up the shipping cranes like ants fleeing flooding nests. The cranes are the safest places to be; they rise more than a hundred feet above the docks. Survival means reaching a high place—a crane, a bridge, or, in Long Beach Harbor, a fifty-foot-high heap of scrap metal on Pier T. The tsunami tide lifts everything that floats, but it can't budge the scrap heap, which becomes an island crowded with survivors.

The tsunami reaches its highest level in the low-lying neighborhoods of San Pedro, Wilmington, and Long Beach and begins to flow back like a wave down the face of a beach. The wrack line tracks, more or less, a contour line about twenty feet above sea level. The damage covers more than twenty square miles and includes the entire seaport complex, nearby marinas, and parts of several neighborhoods. To the east and south, the wave has laid waste to another twenty square miles of Seal Beach, Sunset Beach, Huntington Beach, and Newport Beach. Over the next hour, several more waves wash in, adding to the damage.

TSUNAMI PHYSICS AND HISTORY

Tsunamis are waves triggered by sudden displacements of ocean water. Their causes, from most common to least, are undersea fault movements during earthquakes, landslides, undersea volcanic eruptions, and large meteorite impacts. The largest known tsunami in Earth history happened 65.5 million years ago, when a six-mile-wide meteorite splashed down near what is now Mexico's Yucatan Peninsula, unleashing devastation that killed roughly two-thirds of all species on Earth, including the dinosaurs. (The tsunami itself didn't cause the extinction; global climate change from dust kicked up by the impact did that.) Geologic evidence shows that the wave washed inland for many miles along the Gulf of Mexico coastline; in so doing, it surely took out many dinosaurs. The tsunami was probably more than a half-mile high at the impact site. Although the wave would have lowered rapidly as it radiated away from the impact site, I would bet that some unlucky dinosaur in what is now Mexico holds the world record for the tallest surfed wave. (We'll come back to surfing in chapter 5.)

The largest tsunami in human history, measured by run-up, took place on July 9, 1958, in Lituya Bay, Alaska, about 125 miles west of Juneau. An earthquake triggered a massive rockslide that plopped into the northern arm of the bay, creating a splash that rose 1,700 vertical feet (more than a quarter-mile) up the opposite ridge. The wave surged down the bay toward the ocean, flattening the forest several hundred feet above sea level, forming a still-visible scar called the *trim line* (figure 2.2). Three boats lay anchored in the bay at the time. The *Sunmore* capsized and sank, killing both people aboard. The *Edrie* washed up onto land with all hands safe. The forty-four-foot *Badger* went surfing. The tsunami snapped the boat from its anchor chain and took it on a ride *up and over the forest canopy*. The two people on board reported looking down on the tops of hundred-foot-tall trees as the wave swept the boat out of the bay and dumped it into the Pacific. They were able to launch their small skiff as the boat foundered and were later rescued. Geologists studying growth rings on wave-damaged trees around the bay have found evidence for at least three big tsunamis prior to the 1958 wave (a 395-foot wave in 1853, a 200-footer in 1899, and a 490-footer in 1936). These discoveries forced a rethinking about the importance of landslides in triggering tsunamis and led, in no small way, to our current thinking about tsunami hazards in Southern California.

FIGURE 2.2. The world's biggest splash—at Lituya Bay, southeastern Alaska, on July 9, 1958—testifies to the power of landslides to make tsunamis. The landslide, triggered by a large earthquake near the head of the bay, heaved a wave more than seventeen hundred feet up the opposite shore (at the point marked "maximum wave height" on the photo). As the wave surged west toward the ocean, it leveled the forest several hundred feet above sea level, creating a scar known as the *trim line.* A forest of opportunistic alders sprang up in the wrecked zone. (Photographed by Charles and Emma Jean Mader on October 1, 1997, thirty-nine years after the event.)

As I explained in chapter 1, side-by-side fault movements dominate the plate-tectonic setting of Southern California. That's unlike most of the Pacific Rim, where subduction dominates. Subduction—the process in which an oceanic plate dives into the planet at an ocean trench—generates most of the world's large earthquakes and active volcanoes. Subduction zones ring the Pacific Ocean, feeding magma to the volcanoes of the Pacific Ring of Fire (figure 2.3). Subduction sometimes makes tsunamis, and the reason, we think, goes like this: As a plate of ocean floor subducts, it sticks to the edge of the plate above, bending it gradually downward over many years. When the down-bent plate snaps free during an earthquake, it pops upward and displaces the ocean water above it to make a tsunami. When we plot the world's tsunamis on a map, we find that most of them line up with subduction zones. The Pacific Ocean has the greatest total length of subduction zones, and so most of the world's tsunamis are spawned in the Pacific.

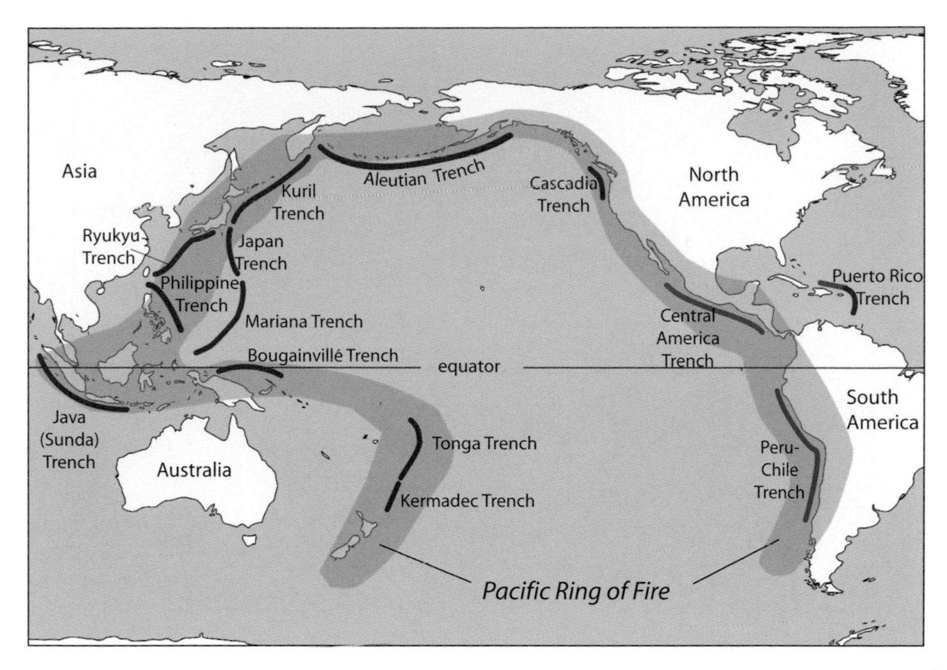

FIGURE 2.3. The Pacific "Ring of Fire" spawns most of the world's volcanoes, earthquakes, and tsunamis. The dark lines mark ocean trenches, also known as *subduction zones.* Subduction of oceanic plates down these trenches creates both volcanoes and undersea earthquakes, some of which form tsunamis. For a more detailed view, see the Earth's Tectonic Plates map at the front of the book.

There's no subduction happening offshore anywhere near Southern California today (although there once was; in chapter 4, I'll tell you the story of how and why that changed). Instead, throughout the Continental Borderland, the dominant movement of seafloor faults is side-by-side sliding. Side-by-side fault movements are less likely to make tsunamis than up–down fault movements, for the same reason that your hand, held flat in a tub of water, will displace less water if you move it side-to-side than if you move it up or down. Yet up–down fault movements do occur in the Borderland; if they didn't, we wouldn't have islands and shallow banks that rise thousands of feet above intervening deep basins. Up–down movements often happen where side-by-side moving faults *bend.* Few faults run straight for their entire length. Depending on which way a fault bends, the rocks along the bend may either pull apart, causing them to sag down and form a basin, or crunch together, causing them to rise up into an island at sea or a mountain on land. These two types of bends, illustrated in figure 2.4a, are known,

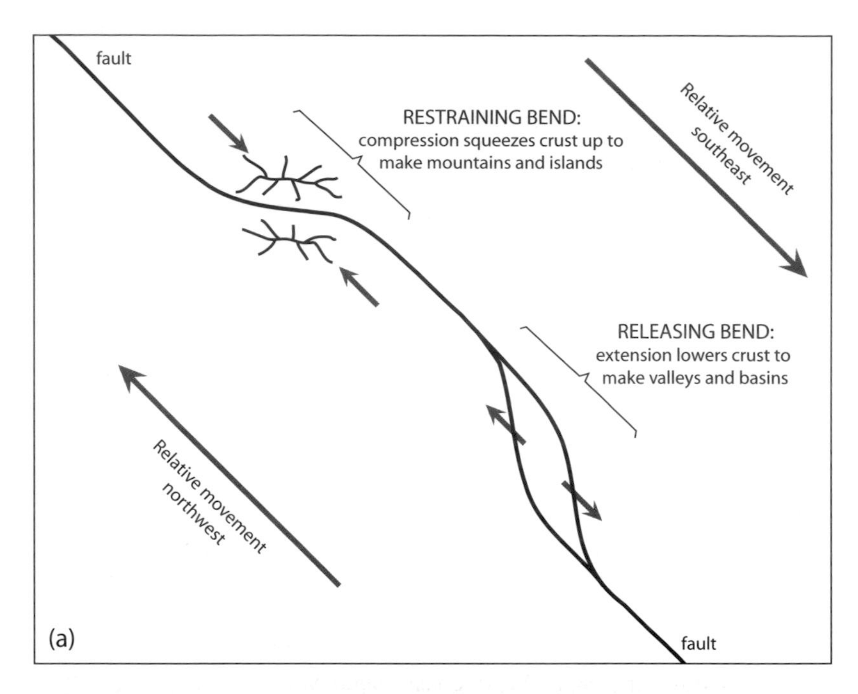

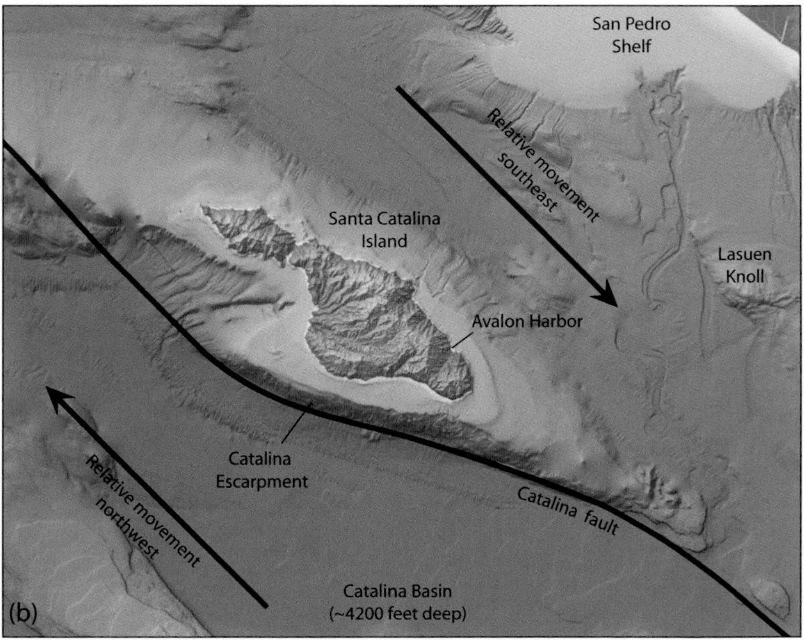

FIGURE 2.4. How bends in side-by-side shifting faults can create up or down movements of the crust. (a) Releasing bends cause the rocks on either side of the fault to separate, forming valleys and basins. Restraining bends, by contrast, cause the rocks on either side of the fault to push into each other, squeezing up mountains and islands. (b) Santa Catalina Island has risen along a large restraining bend in the Catalina fault. (NOAA shaded relief image, with labels added.)

respectively, as *releasing bends* and *restraining bends*. Santa Catalina Island, for instance, has been hoisted out of the sea by compression along a large restraining bend in the Catalina fault (figure 2.4b). The largest example of a restraining bend in California is the Big Bend in the San Andreas fault, which has pushed up the Transverse Ranges along the Big Squeeze (see figure 1.3).

Given that up–down fault movements clearly occur in the Continental Borderland, what does that say about tsunami risk there? The Catalina fault has received the most attention, because of its large restraining bend and its proximity to the mainland. Computer modeling suggests that an earthquake of magnitude 7.6 (the high end of likely magnitude for the Catalina fault) could generate a tsunami with six to twelve feet of run-up along parts of the Los Angeles and Orange County coastlines. (Note: This is for a tsunami triggered by seabed fault displacement alone. Tsunamis caused by large undersea landslides are likely to be bigger, as I'll explain in a moment.) Likewise, displacement on the Coronado Bank fault west of San Diego (maximum likely magnitude 7.5) might hurl six to twelve feet of run-up onto the San Diego coast. Such waves would certainly be destructive. But six to twelve feet of run-up pales in comparison to the run-ups from large subduction-generated tsunamis. The two biggest tsunamis so far this century, both subduction-generated, had run-ups of *more than 100 feet* on nearby coasts.* Clearly, the absence of subduction nearby is a good thing when it comes to tsunami risk in Southern California.

What about tsunamis that travel to Southern California from distant subduction zones? Here there is little cause for worry. The nearest active subduction zones—the Cascadia Trench and Central America Trench (figure 2.3)—are, respectively, more than five hundred and twelve hundred miles away. Moreover, the maze of islands and shallow banks that make up the Continental Borderland are likely to dampen far-traveled tsunamis as they enter the Southern California Bight. History bears this out. The five biggest Pacific earthquakes of the past hundred years (all subduction-generated) each triggered huge tsunamis along nearby coasts, but all arrived in Southern California highly subdued. Here's a summary of these five great quakes and the tsunamis they spawned.

The Kamchatka earthquake (magnitude 9.0) of November 4, 1952, lashed parts of the Kuril Islands with forty feet of run-up, but the waves

* The Indian Ocean tsunami of December 26, 2004, had run-ups of 115 feet along the coast of Sumatra, and the Japanese tsunami of March 11, 2011, had run-ups of more than 100 feet in parts of Iwate Prefecture.

dampened to one or two feet upon reaching Los Angeles and San Diego. Waves from the great Alaskan earthquake (magnitude 9.2) of March 28, 1964, wreaked destruction across coastal Alaska, with more than *two hundred feet* of run-up in Valdez Inlet; but when the waves arrived in Southern California five hours later, they were all of two feet high (although they did damage some boats and piers with rapid currents). I watched for waves from the Chilean earthquake (magnitude 8.8) of February 27, 2010, from atop a bluff near my San Diego home but saw nothing; tide gauge records showed them passing the coast as ghostly pulses less than a foot high—too small to pick out in the local surf (although they did cause strong currents in San Diego Bay). Waves from the Japanese earthquake (magnitude 9.0) of March 11, 2011, arrived on the San Diego coast about one to two feet high, causing powerful currents in bays and harbors but little damage. The greatest damage from a distant tsunami in Southern California so far came from the Chilean earthquake of May 22, 1960—the most powerful quake ever measured by modern instruments (magnitude 9.5). The waves arrived in Southern California thirteen hours after the quake, surging two to six feet along the coast and unleashing fierce currents that ripped boats from moorings and smashed piers in San Diego and Los Angeles harbors. Geophysicists think that the May 1960 Chilean quake was close to the biggest possible on Earth (see text box in chapter 3 for an explanation why). The evidence above points to one conclusion: Even the largest distant tsunamis fade mostly to a whimper upon arrival in Southern California, and cause only moderate damage.

To sum up the evidence of the last few pages: Southern Californians can expect six to twelve feet of run-up from a tsunami caused by seabed fault displacement in the Continental Borderland, and perhaps two to six feet of run-up from the largest distant tsunamis that arrive here from elsewhere in the Pacific. Doesn't sound too bad, does it? There's just one problem—the *landslide* scenario. When a truly devastating tsunami—a wave with, say, twenty feet or more of run-up—eventually surges ashore in Southern California, it will almost certainly come from an undersea landslide.

THE LANDSLIDE SCENARIO

Before exploring landslide-generated tsunamis, let me explain how we know about the extent of undersea landslides and other seabed features in the Continental Borderland. In the past few decades, a new technology called *multibeam bathymetry* has revolutionized our views of the ocean floor. (Bathymetry is the science of mapping seabed depths.) The technology is based on sonar (short for *sound navigation and ranging*), which uses sound pulses called *pings* that echo off the seafloor or objects in the water column. The speed of sound in seawater averages 4,945 feet per second (varying a bit with temperature, salinity, and pressure), so echo time translates to water depth. Sonar technology originated in the early 1900s and expanded greatly during World War II to hunt submarines. (Visualize cringing submariners in wool caps listening for the pings of hunter ships overhead—an iconic image from countless war movies.) In multibeam bathymetry, dozens of ping-emitters are arrayed at slightly different angles beneath a ship, or beneath an instrument towed at depth behind a ship. Multiple pings radiate outward from these emitters, covering a swath of seabed more than thirty miles wide and returning hundreds of simultaneous depth readings every few seconds. The resulting computer-processed images look practically as if you had drained the ocean to look at the seabed with your own eyes. Figure 2.5 shows an example.

Multibeam surveys have revealed dozens of undersea landslides throughout the Continental Borderland. Most are less than one square mile in area—far too small to be a tsunami worry. But two much larger slides—the Goleta Slide west of Santa Barbara and the Palos Verdes Slide southwest of Long Beach—tell a different story. Both were last active several millennia ago, so we don't know how big the resulting tsunamis were. The best we can do is use computer models to mimic the slides and the resulting displacement of ocean water. Whatever models we use, the results are sobering.

The Goleta Slide sprawls across twenty-five square miles of seabed a few miles west of Santa Barbara (figure 2.6). It begins in about three hundred feet of water at a scarp—an immense cliff, more than six hundred feet high, marking where the slide broke away from the edge of the continental shelf. The slide has three distinct lobes that may or may not have moved together. Each lobe is five to eight miles long and ends in water more than eighteen hundred feet deep. The toes of the lobes rise above the surrounding seabed as high as a fifteen-story building. By beaming acoustic waves into the seabed, geologists have detected at least six older slides buried below the present-day Goleta Slide, telling us that landslides have happened here multiple times.

Together, the three lobes of the Goleta Slide contain one-third to one-half a cubic mile of debris—roughly fifty times the volume of Alaska's Lituya Bay landslide, which I described earlier (figure 2.2). To visualize

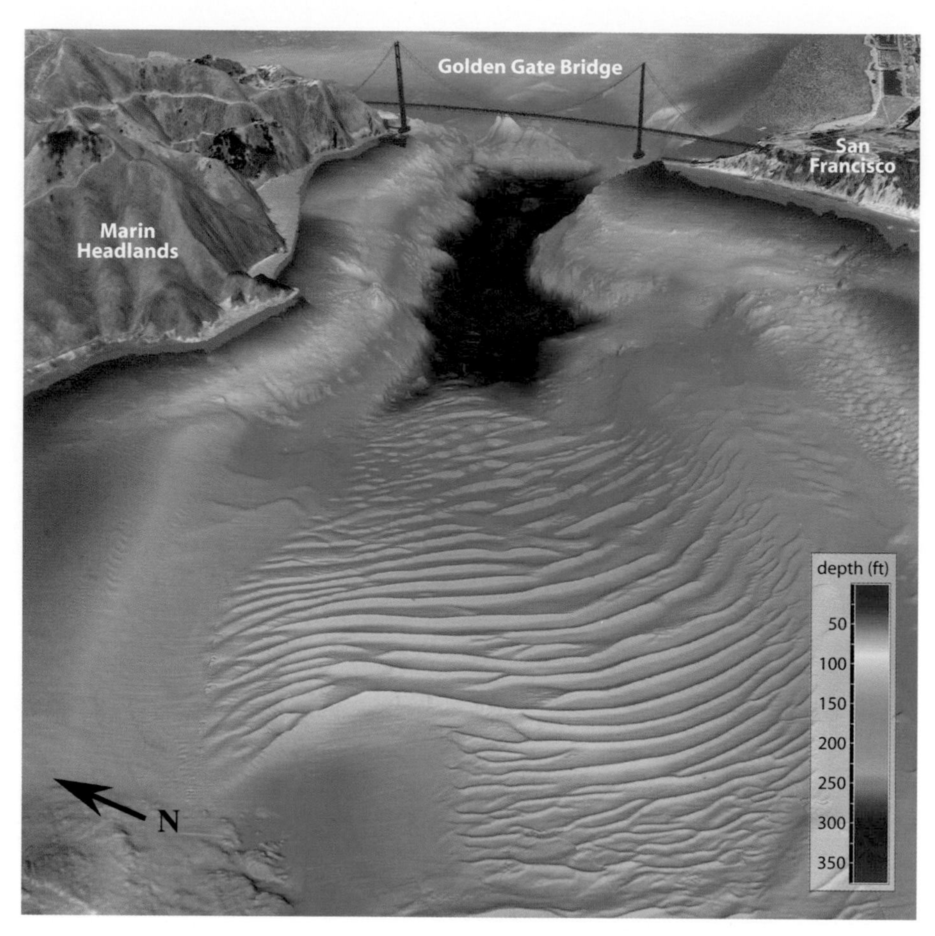

FIGURE 2.5. An example of multibeam bathymetry. Colors reflect depth in this high-resolution image that looks east through the Golden Gate, the entrance to San Francisco Bay. Powerful tides funneling through the narrow strait have scoured the seabed more than three hundred feet deep at the narrows. As the ebbing tides slow in the wider foreground region, they drop sand to form gigantic underwater dunes as much as thirty feet high and seven hundred feet from crest to crest. Traditional navigation charts cannot reveal this level of detail. (U.S. Geological Survey image; see Barnard et al. 2006.)

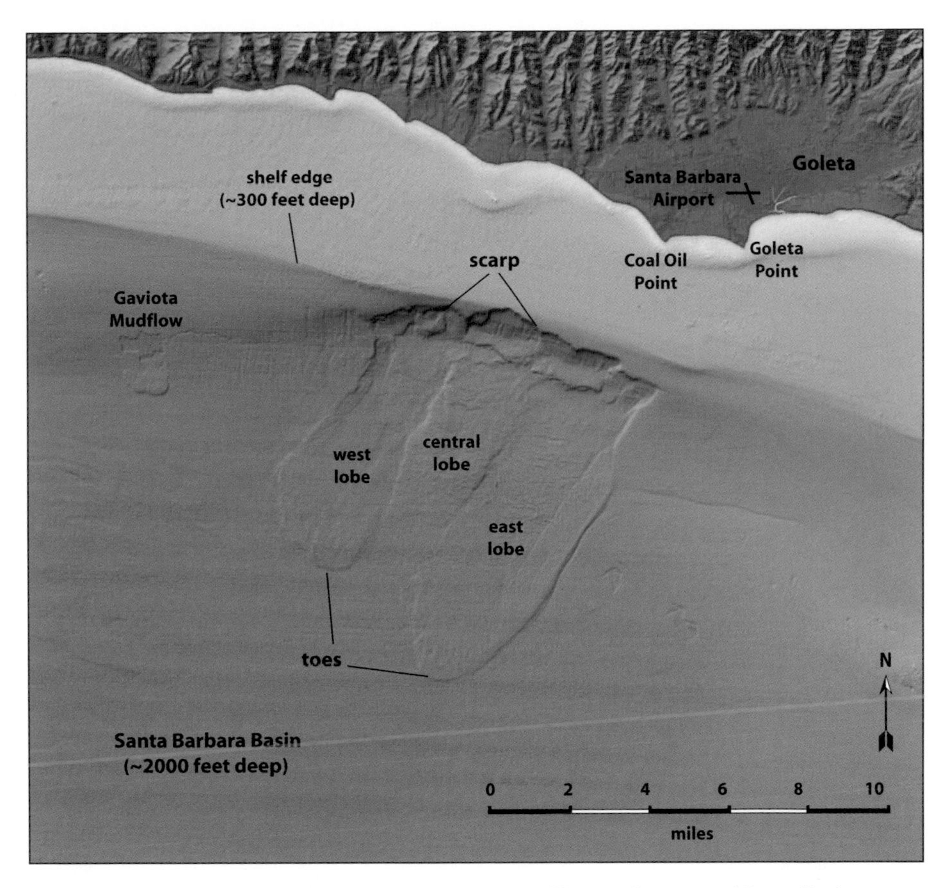

FIGURE 2.6. Multibeam image of the Goleta Slide about fifteen miles west of Santa Barbara. The slide consists of three lobes covering more than twenty-five square miles. (NOAA shaded relief image, with labels added.)

what such a displacement could do in the Santa Barbara Channel, imagine heaving a trash can filled with concrete into an average-sized swimming pool—the volumes are proportionally about the same.* How big would the resulting tsunami have been? It depends on what assumptions you make about lobe simultaneity, slide volume, and slide velocity, but conservative estimates suggest run-ups of more than thirty feet along at

* Here's the math to support the analogy. The Santa Barbara Channel, 80 miles long by 25 miles wide and averaging 700 feet (0.13 miles) deep, contains roughly 250–300 cubic miles of water, or about 500–800 times the volume of the Goleta Slide. A typical household swimming pool holds 2,500 to 3,000 cubic feet, which is roughly 500–700 times more than a typical 32-gallon (4.3 cubic feet) household trash can.

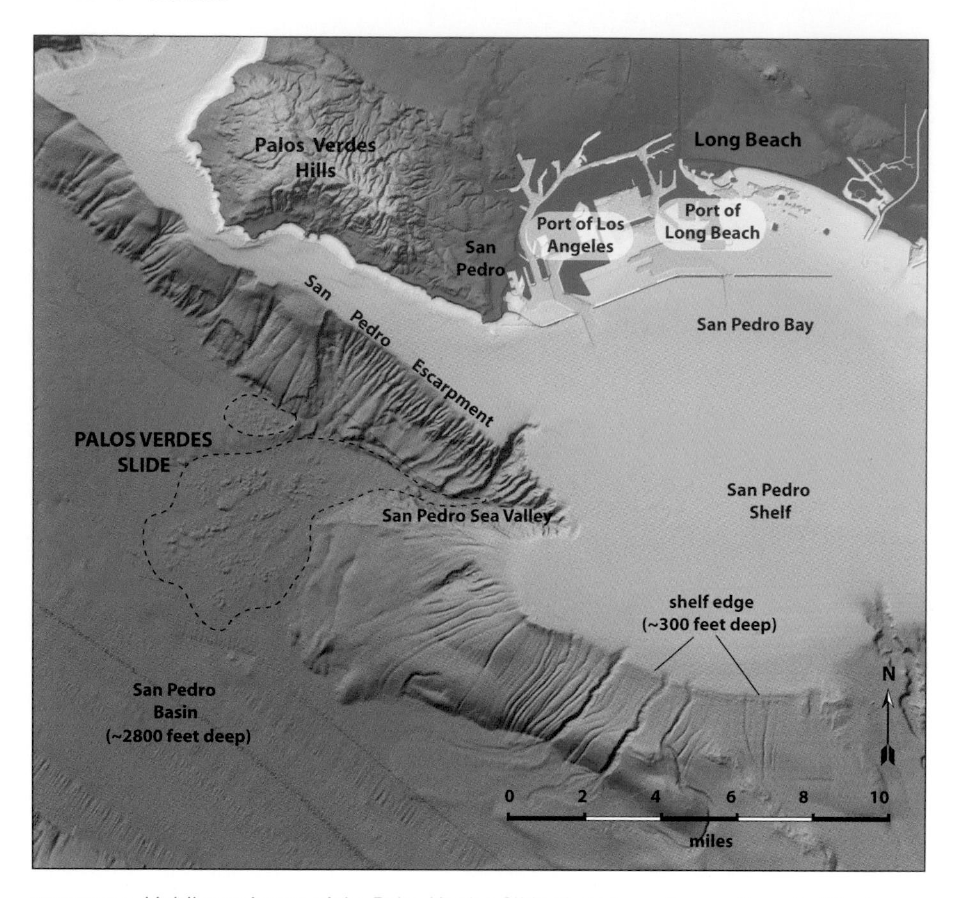

FIGURE 2.7. Multibeam image of the Palos Verdes Slide about ten miles southwest of the Los Angeles and Long Beach seaports. The slide cascaded down the San Pedro Sea Valley about seventy-five hundred years ago, almost certainly triggering a large tsunami. A repeat of such an event would probably destroy the seaports and large portions of nearby coastal cities. (NOAA shaded relief image, with labels added.)

least twenty miles of coastline in Santa Barbara County. The worst-case scenario, in which the entire Goleta complex moved at once, might have produced run-ups of more than sixty feet along much of the fifty-mile stretch of coast between Point Conception and Carpinteria. Were that to occur today, it would wipe out large portions of Santa Barbara, Ventura, and other coastal communities.

As bad as that may seem, a larger potential tsunami disaster may await a few miles offshore of the Palos Verdes Peninsula, where the ocean floor plunges more than two thousand feet down the San Pedro

Escarpment to the floor of the San Pedro Basin. The slopes here are among the steepest anywhere in the Continental Borderland. Several big canyons notch the escarpment, the largest of which is the San Pedro Sea Valley. Disgorged from the valley's mouth in 2,600 feet of water lies the Palos Verdes Slide. Unlike the Goleta Slide, whose lobes may have moved separately, the Palos Verdes Slide was a single colossal landslide that cascaded down into the San Pedro Basin about 7,500 years ago. The slide had enough momentum, upon reaching the basin floor, to carry house-sized blocks of rock up to five miles across the nearly flat seabed. Multibeam images reveal debris from the slide spread across fifteen square miles of the San Pedro Basin (figure 2.7).

At about one-sixth of a cubic mile in volume, the Palos Verdes Slide is smaller than the Goleta Slide. But it's still twenty times bigger than the Lituya Bay landslide, and—more importantly—it's much closer than the Goleta Slide to the densely urbanized Los Angeles coastline. Using seismic reflection, geologists have discovered at least six, and possibly nine, older landslides buried below the present Palos Verdes Slide, telling us that submarine landslides here—like those at Goleta—are repeat events. I based my fictional tsunami at the start of this chapter on a repeat performance of the Palos Verdes Slide, giving free reign to imagination but not to exaggeration. The twenty-foot run-up that I used for that scenario falls squarely within the range of estimates in the scientific literature for a repeat of the Palos Verdes Slide.

Southern California's geologic stew of active faults and undersea landslides guarantees that a large tsunami will wash ashore here again one day. But in any given year, the odds of a devastating tsunami are considerably lower than those of a large earthquake—the topic of the next chapter.

3

Earthquakes

The formula for a happy marriage? It's the same as the one for living in California: when you find a fault, don't dwell on it.

—Jay Trachman, radio humorist

Although not large compared to other California ranges, the Santa Ynez and Santa Monica mountains are some of the steepest and most spectacular in California. It's because they are still being born. Rock layers less than one million years old, originally flat-lying, have been pushed and tilted to crazy angles throughout the still-rising ranges. Take a phone book (remember those?), lay it flat on a table, and push it together with your two hands so that the pages arch upward in the middle (forming, in geologic terminology, an *anticline*). That mimics, in broad form, the rise of the Santa Ynez and Santa Monica mountain ranges, both of which are arching upward, earthquake by earthquake, to escape the vise of Southern California's Big Squeeze (see chapter 1). The mountains are younger, even, than some of the rivers that flow through them. Malibu Creek, for instance, probably flowed to the ocean before the Santa Monica Mountains existed. As the mountains squeezed upward, the creek maintained its course, slicing into the rising land like a stationary saw blade through a rising log. The result is Malibu Canyon. The Ventura River and its tributaries have done the same, slicing deep canyons into the still-rising Santa Ynez Mountains.

When most of us think of earthquakes, visions of destruction come to mind. But it's worth considering what earthquakes have given us before exploring what they can—at any moment and without warning—take away. Earthquakes built Southern California's geography. Every quake that rattles the region today adds to the work of numberless predecessors by hoisting a mountain range or an island, or dropping a valley or undersea basin, bit by bit. Erosion adds its mark, sculpting topography raised by earthquakes. Without earthquakes, erosion wouldn't have much to do.

Earthquakes also brought oil wealth to Southern California, particularly to the Big Squeeze region, as I'm reminded whenever I see the oil platforms in the Santa Barbara Channel or pass pump jacks bobbing on the hills near Ventura. Rock layers caught in the Big Squeeze respond to the crushing forces there by breaking along faults and by bending into what we call *folds*. (Take the flat-lying blankets on your bed and push them aside; the wrinkles you see give a fair image of what the folded rock layers within the Big Squeeze look like underground.) As it turns out, the earthquake-broken and earthquake-bent rock layers in the Big Squeeze create ideal oil traps. To see why, let me make a quick aside.

Most crude oil begins where plankton—tiny floating ocean plants and animals—die and fall from surface waters to the seabed, where they pile up to form carbon-rich sediments. If this carbon-rich glop is buried deep enough, pressure and heat cook it into crude oil. As the Continental Borderland formed during the past twenty million years (a story we'll explore in chapter 4), numerous deep basins stretched open on the seabed. Plankton, particularly diatoms—plant-like microorganisms with tiny silica shells—rained down abundantly into these basins, creating a rock unit known as the Monterey Formation. In its purest state, the Monterey Formation is pure *diatomite*—a powdery, porous, snow-white rock made almost entirely of diatom remains. In areas where the Monterey Formation was buried deeply enough, heat and pressure transformed the diatom-rich organic matter into crude oil. Energy analysts estimate that the Monterey Formation holds fifteen billion barrels of recoverable crude—more than double what the United States currently consumes each year.

Once oil forms underground, it rarely stays put. Oil is, of course, much less dense than rock, so wherever it forms, it always migrates upward. If oil migrates all the way to the surface, it leaks out as natural seeps. At Carpinteria, the coastal bluffs bleed oil tar like Kilauea oozes lava, and you can scarcely walk the beach without stepping on tar blobs. A big natural seep forms the famous fossil trap at La Brea Tar Pits in downtown Los Angeles (*brea* = "tar" in Spanish). If all crude oil escaped via natural seeps, we would not have a modern industrial society. Happily, a lot of oil gets trapped underground. There it may sit for millions of years until someone finds it, drills down to it, and pumps it out. Faults form excellent crude oil traps. If a fault happens to thrust impermeable rock up and over porous rock—as commonly occurs in the Big Squeeze—oil and gas will rise up through the porous rock until they bump into the impermeable rock above. Unable to rise farther, the oil and gas collect there to form a reservoir. Oil and gas also collect where rock layers arch upward in anticlines. The oil migrates up through porous rock layers (such as beds of sandstone) within the anticline until it runs into an impermeable layer, like a bed of dense shale. There the rising drops of oil collect like helium balloons against a ceiling. Billions of barrels of oil, and trillions of cubic feet of natural gas, have been extracted from Southern California's underground architecture of bent and broken rock created by earthquakes.

So much for the good that earthquakes may do. Now for the bad.

SHAKE, RATTLE, AND ROLL

By the time you finish this chapter, the odds are good that at least one Southern California fault will have snapped with a detectable earthquake. You can go check.* Much of Southern California experiences a ceaseless drumroll of tiny quakes, reflecting the Pacific Plate's ongoing efforts to drag us toward Alaska. Large earthquakes are, of course, much rarer than small ones. The rare big quakes that punctuate the background crackle of smaller ones represent the larger victories in the Pacific Plate's relocation campaign for Southern California. Yet big quakes, while comparatively rare, are naturally the ones we worry about. A recent U.S. Geological Survey estimate puts the cost of a very large Southern California earthquake at over $200 billion, with at least eighteen hundred fatalities and fifty thousand seriously injured.

Most of the world's earthquakes occur where tectonic plates grind against each other. That's why a world map of earthquakes translates directly into a map of the Earth's plates (figure 3.1). Where plates rub, friction locks them together. As the stuck plates try to move, stress builds along faults. The rocks on each side of a fault bend to take up the strain—but eventually they reach their breaking point. The fault snaps, the rocks leap to a new position, and the energy released surges outward in pulsing spasms called *earthquake waves*. People caught in big

* To see the most recent earthquakes, updated hourly, go to the website of the Southern California Earthquake Center, located (as of this writing) at www.data.scec.org/recent/.

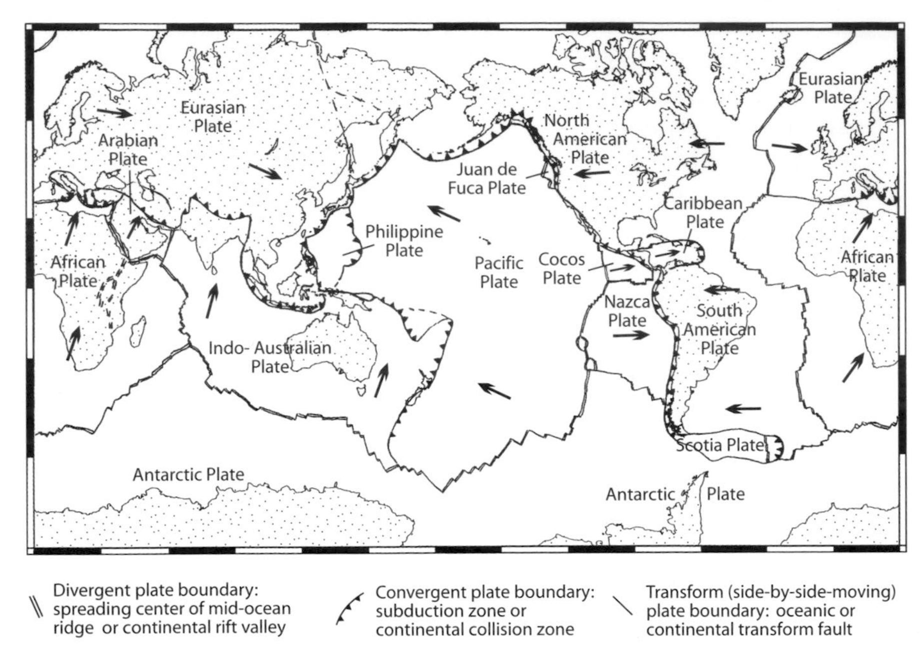

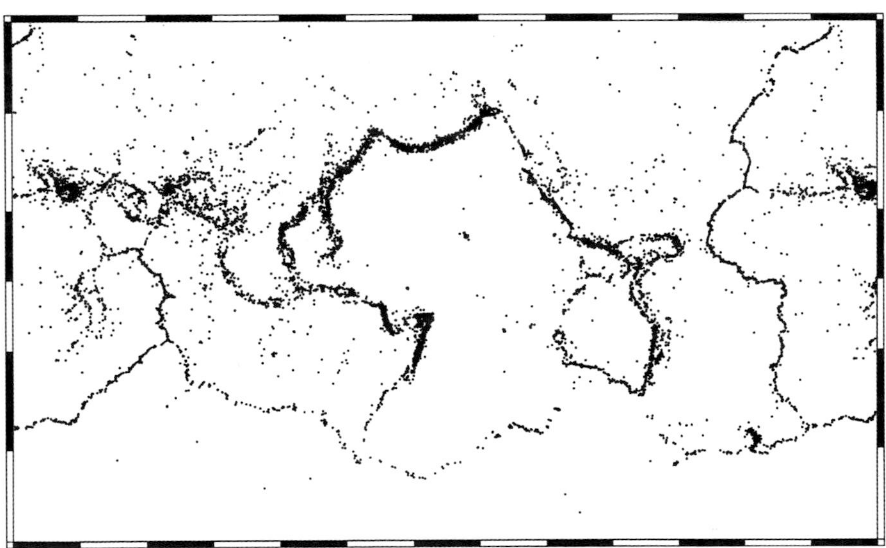

FIGURE 3.1. The upper map shows the Earth's major tectonic plates, with arrows indicating the direction of plate movement. (For a more detailed view, see the Earth's Tectonic Plates map at the front of the book.) The lower map shows global earthquakes recorded over a two-decade period. You can see that most earthquakes line up with plate edges.

quakes describe the ground moving like ocean swells. It was "like we were on a surfboard in the backyard" one San Diegan reported after the April 4, 2010, Mexicali earthquake. The motion "produced a sensation of giddiness, and some were sick similar to sea sickness," wrote one witness to the historic 1857 Fort Tejon earthquake, adding, "everything (houses, trees, cattle and people) had the appearance of being drunk."

Figure 3.2 portrays current forecasts for the intensity of ground shaking and damage from earthquakes in Southern California. Not surprisingly, the San Andreas fault is marked as having the highest risk of violent shaking along its whole length. Looking at the coastal counties, you can see that the map sends a clear message: San Diego County is the safest place to be. The Orange County coast has a bit higher risk, Los Angeles County is higher still, and coastal Ventura and Santa Barbara counties are earthquake nightmares waiting to happen. What's going on to create these differences? It's not as if coastal San Diego doesn't have active faults and a history of large earthquakes. Indeed, no place along the Southern California coast is far from an active fault. The different pictures of coastal risk painted in figure 3.2 come down, at least in part, to the big gorilla of California's faults—the San Andreas.

To understand why, it helps to understand how geologists figure out those relative risk levels shown in figure 3.2. The approach integrates six types of information. First, fault length: Other things being equal, longer faults are capable of great slip and, therefore, stronger quakes. Second, earthquake history: Faults that have shifted violently in the past are more likely to shift again—especially if an unusually long time has passed since the last big quake. Third, slip rates: Faults that have had higher rates of movement in the past are more likely to lurch in the future. Fourth, ground deformation: The more the ground along a fault bends and distorts from year to year, as measured by precision surveying,* the more likely the fault is accumulating strain that will lead it to snap with another quake. Fifth, rupture direction: When a

* Two surveying methods based on satellite technology, GPS and InSAR, allow us to measure ground movements with fraction-of-an-inch precision over large areas, which is a key step for estimating how much strain is building along active faults. The global positioning system (GPS) uses a constellation of twenty-four U.S. military satellites that beam down precisely timed radio signals. A GPS receiver triangulates its position using signals received simultaneously from several satellites. Interferometric synthetic aperture radar (InSAR) uses reflections from microwaves beamed down to the Earth's surface to develop maps of surface change. Repeated satellite flyovers reveal the movements of ice sheets, volcanoes swelling from migrating magma, and the ground shifting before and after earthquakes.

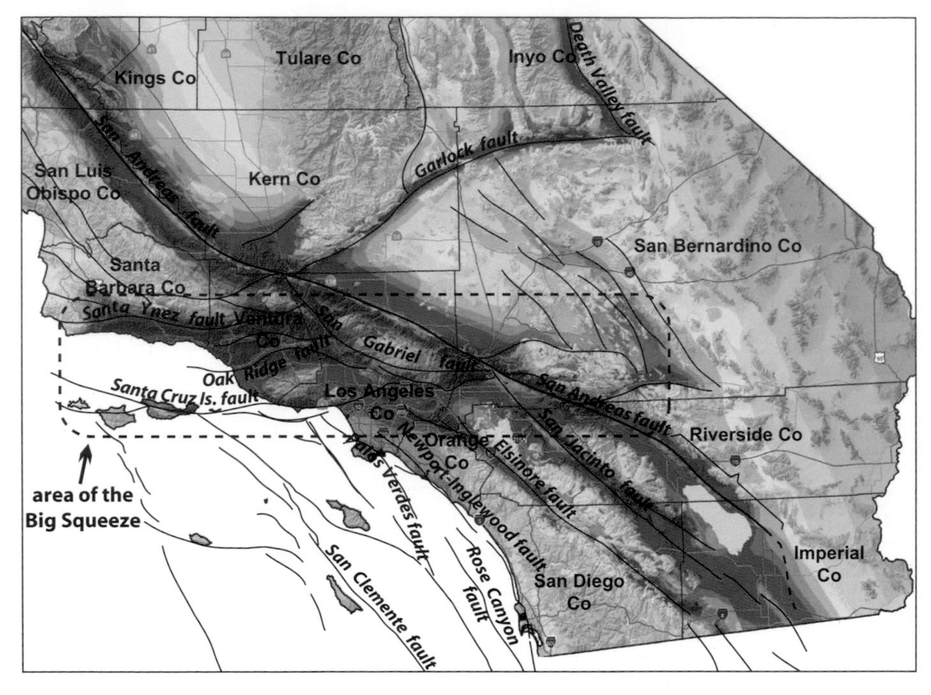

FIGURE 3.2. Map of predicted earthquake shaking potential in Southern California, developed jointly by the California Geological Survey and the U.S. Geological Survey. Warmer colors represent higher risk of violent shaking. Of the five coastal counties in Southern California, San Diego County ranks as lowest risk, while Los Angeles, Ventura, and Santa Barbara counties rank as highest risk. (From Branum et al. 2008, with faults and labels added.)

fault shifts, the rupture tends to travel along the fault like a fast zipper, with worse shaking in the direction that the fault "unzips." Finally, underlying geology: Hard, dense rocks such as granite and other bedrock transmit seismic waves efficiently and therefore shake less. But loose materials like sand or uncompacted landfill are inefficient transmitters of seismic energy. Earthquake waves slow down in these materials, and that boosts their amplitudes and causes more shaking. Moreover, loose materials are also more prone to landslides and to liquefaction, in which the ground turns into a weak, jelly-like mass.

If we take those six concepts above and apply them to the San Andreas fault, we find that the *southernmost* section of the fault appears to be at the greatest risk for uncorking a great earthquake. Ask any seismologist which part of the San Andreas fault he or she worries about the most, and the answer will probably be the southernmost section—specifically the segment from the Salton Sea to the Transverse Ranges. Notice in

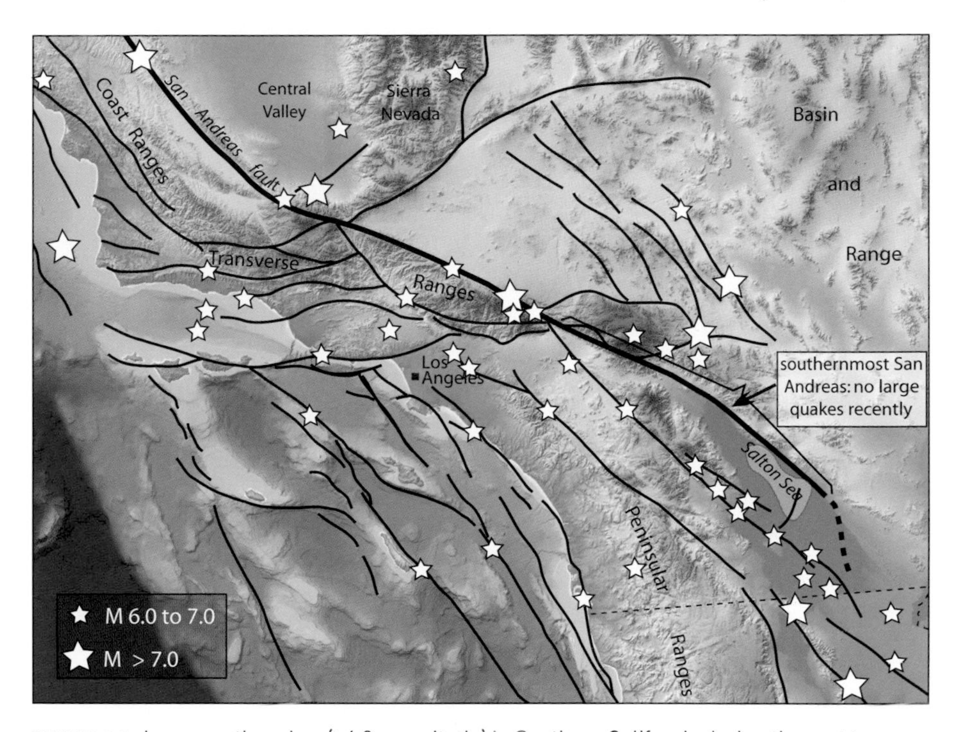

FIGURE 3.3. Large earthquakes (≥6.0 magnitude) in Southern California during the past two hundred years. Many faults have had at least one large quake during this time. A worrisome exception is the southernmost part of the San Andreas fault, which has not fired off a large quake since about 1680. When this section of the fault next ruptures, it seems likely to hurl destruction across much of Southern California, as illustrated in figure 3.5. (Shaded relief image from NOAA, with labels added; quake locations and magnitudes from the Southern California Earthquake Center.)

figure 3.3 that there have been no large historical quakes along this section of the fault. Geologic evidence indicates that the last major earthquake here (estimated at magnitude 7.8; see text box for an explanation of earthquake magnitude) was in roughly the year 1680, or more than 330 years ago. (That date of 1680 is based on disrupted sedimentary layers along the fault, not human records, although resident native tribes were undoubtedly thrashed by the quake.) This 330-year-plus time-gap since a big quake on the southernmost San Andreas is troubling because before that 1680 quake, the southernmost San Andreas seems to have ruptured with an estimated M7 or greater quake about every 180 to 200 years. Earthquakes don't repeat like clockwork—not even close—but many faults seem to rupture on a rough cycle. Regular earthquakes are a bit like regular bowel movements; they relieve accumulating

Measuring the Strength of Earthquakes

The *moment magnitude scale* (M-scale) is the standard currently used to express the strength of earthquakes, supplanting the older *Richter scale* that dominated earthquake measurement for much of the twentieth century. Although the two scales give similar values, the M-scale is more accurate because it uses a wider range of variables, including

- vibrations from several types of seismic waves received by seismographs (instruments that measure ground motion during earthquakes)
- the strength of the rock that ruptures
- the total area of the fault that moves
- the total displacement of the fault

The M-scale is logarithmic, so that each increase of 1.0 represents ten times more seismic wave amplitude and roughly thirty times more seismic energy released. The largest earthquake ever recorded by modern instruments—on May 22, 1960, off the coast of Chile—equaled M9.5. The strength of rock and the amount of fault displacement put an upper limit on the possible magnitude of any earthquake. The longer the fault, the greater the displacement can be when the rock breaks—and thus the bigger the quake. The stronger the rock (strength being related mostly to temperature and mineral composition), the more strain energy the rock can store as it bends before breaking—and thus the bigger the quake. But because faults have finite length and rocks have finite strength, we think that magnitude M10 is close to the upper limit possible for earthquakes on Earth, although other planets could

pressure and allow materials to keep moving in the direction they need to go. Not to be crude about it, but the southernmost San Andreas fault appears to be seriously constipated and in need of a Big One.

Measurements of ground deformation back this up. Precision surveying (see footnote on p.47) shows that the area west of the San Andreas fault (between the San Andreas and San Jacinto faults) is creeping northwest at nearly a half-inch per year in relation to the east side. Since this movement isn't causing earthquakes, the rock must be *bending* to allow the movement to happen. Just as you can bend a stick only so much until it snaps, rock along a fault can bend only so far before it snaps and leaps to a new position, slinging forth its pent-up energy as an earthquake. The more the rock bends before breaking, the bigger the quake when it finally breaks. When the southernmost San Andreas eventually ruptures, the quake is likely to be at least M7.8, similar to the 1680 theoretically have larger quakes. The largest historical earthquakes in California (see figure 3.4) have been M7.8 to M7.9.

The moment magnitude scale measures the energy released by an earthquake, but it says little about the quake's effects on people and infrastructure. For this, you'll more often see the *Mercalli intensity scale,* developed more than a century ago by the Italian seismologist Giuseppe Mercalli. The Mercalli scale characterizes the strength of an earthquake by the amount of shaking and damage. The scale uses Roman numerals I through XII. Some examples:

II—Felt by a few stationary people, especially in the upper floors of buildings; suspended objects like lamps may swing.

VII—People are frightened and run outside; plaster walls crack, windows break, some chimneys topple, some unstable furniture overturns; weak buildings may sustain considerable damage.

XII—Earthquake waves cause visible undulations of the ground; objects are thrown up off the ground; complete destruction of buildings and bridges of all types.

A given earthquake can have only one moment magnitude value but will have a range of Mercalli intensity values, depending on the distance from the epicenter and the type of rock. A contour map of Mercalli values typically looks like a rough bull's-eye with the highest values near the epicenter, but not always. For example, homes with foundations set in granite ten miles from the epicenter may endure considerably less shaking and damage than homes twenty miles from the epicenter built on loose sand and gravel.

quake and to the two biggest historical quakes along other parts of the great fault: the 1857 Fort Tejon quake (M7.9) and the 1906 San Francisco quake (M7.8; see figure 3.4).

How bad would an M7.8 on the southernmost San Andreas fault be? Researchers affiliated with the Southern California Earthquake Center have used supercomputers to model several scenarios. Their results are presented in beautiful, terrifying animations of Southern California being whipsawed by virtual seismic waves.* The various models disagree in detail, but they agree in two key ways.

First, the direction of rupture—whether the fault "unzips" south-to-north or north-to-south—controls how the damage will be distributed.

* The visualizations are available (as of this writing) at the San Diego Supercomputer Center website: http://visservices.sdsc.edu/projects/scec/.

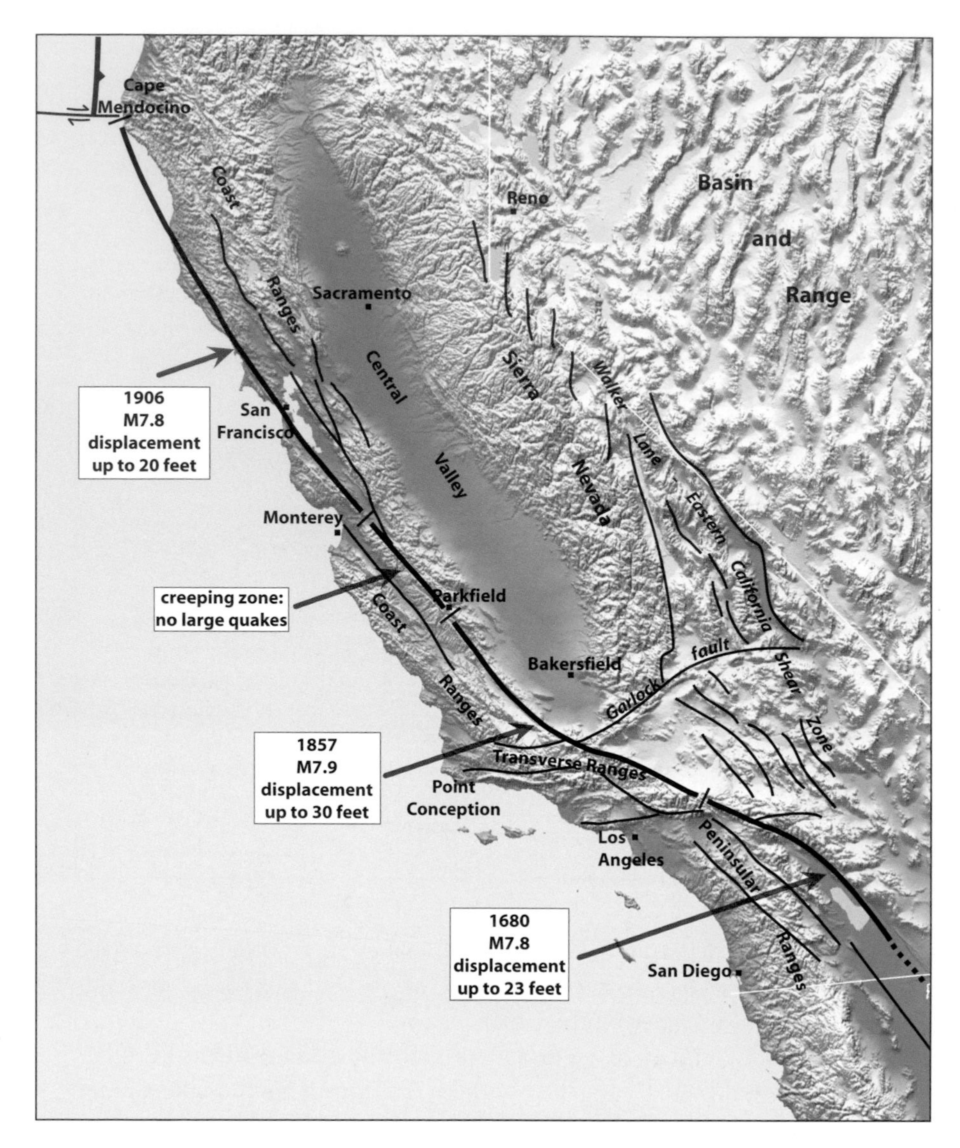

FIGURE 3.4. Two of California's three largest historical earthquakes have occurred on the San Andreas fault: the 1857 Fort Tejon quake and the 1906 San Francisco quake. (The third was the 1872 Owens Valley earthquake, which occurred along the Sierra Nevada Frontal Fault System.) The bracketed areas show the sections of the San Andreas fault that shifted during these events, and the labels indicate the magnitudes and the amount that the fault shifted in feet. No quake of comparable size has occurred on the southernmost section of the San Andreas since about 1680. This section seems likely to rip loose with an earthquake of comparable magnitude in the near future. (Shaded relief image by NASA, with labels added.)

Everyone living in the southern Central Valley or Los Angeles should hope for north-to-south rupture, which will sling most of the seismic energy toward the Imperial Valley and adjacent Mexico (although LA would still get thrashed). But Imperial Valley residents may have the odds in their favor; a more likely scenario may be south-to-north rupture, which will send most of the energy toward Los Angeles. Figure 3.5 shows a simulation of a south-to-north rupture.

A second point of agreement among the computer models is that the basins—the valleys filled with loose sand and gravel eroded from the mountains—will experience some of the most violent and prolonged shaking. The bottom image of figure 3.5 makes this point. Hemmed in by mountains, the seismic waves ricochet back and forth across the Los Angeles and Ventura basins, shaking them severely for several minutes even after the rupturing has stopped. The scene reminds me of a swimming pool after a big kid does a cannonball; the waves bounce off the sides of the pool and cross back and forth through each other, forming a chaos of peaks and valleys. Visualize a toy boat in that pool—that's a house or office building in Los Angeles after an M7.8 quake rips loose along the southern San Andreas fault.

EARTHQUAKE RISK IN THE COASTAL ZONE

Let me now come back to the Southern California coast and address a puzzle that I brought up several pages back: Why does coastal San Diego County have lower earthquake risk than Southern California's other coastal counties, as shown in figure 3.2? A partial answer comes out of the San Andreas "Big One" scenario described above. No matter which direction the San Andreas unzips—south-to-north or north-to-south—the bow wave of advancing seismic energy spreads and dampens by the time it hits San Diego (figure 3.5). Moreover, much of urban coastal San Diego is built on moderately strong sedimentary rock, whereas large parts of urban coastal Orange, Los Angeles, and Ventura counties are built on loose sand and gravel that fill the basins between bedrock mountains. But this isn't the whole story; after all, much of coastal Santa Barbara County is built on strong bedrock, yet it glares high-risk red in figure 3.2.

If you think back to chapter 1 or to the start of this chapter, the higher quake risk in Santa Barbara, Ventura, and Los Angeles counties may come as little surprise. These counties all lie within the Big Squeeze, where rocks being dragged northwest with the Pacific Plate mash against the Big Bend in the San Andreas fault. Frequent earthquakes are the expected

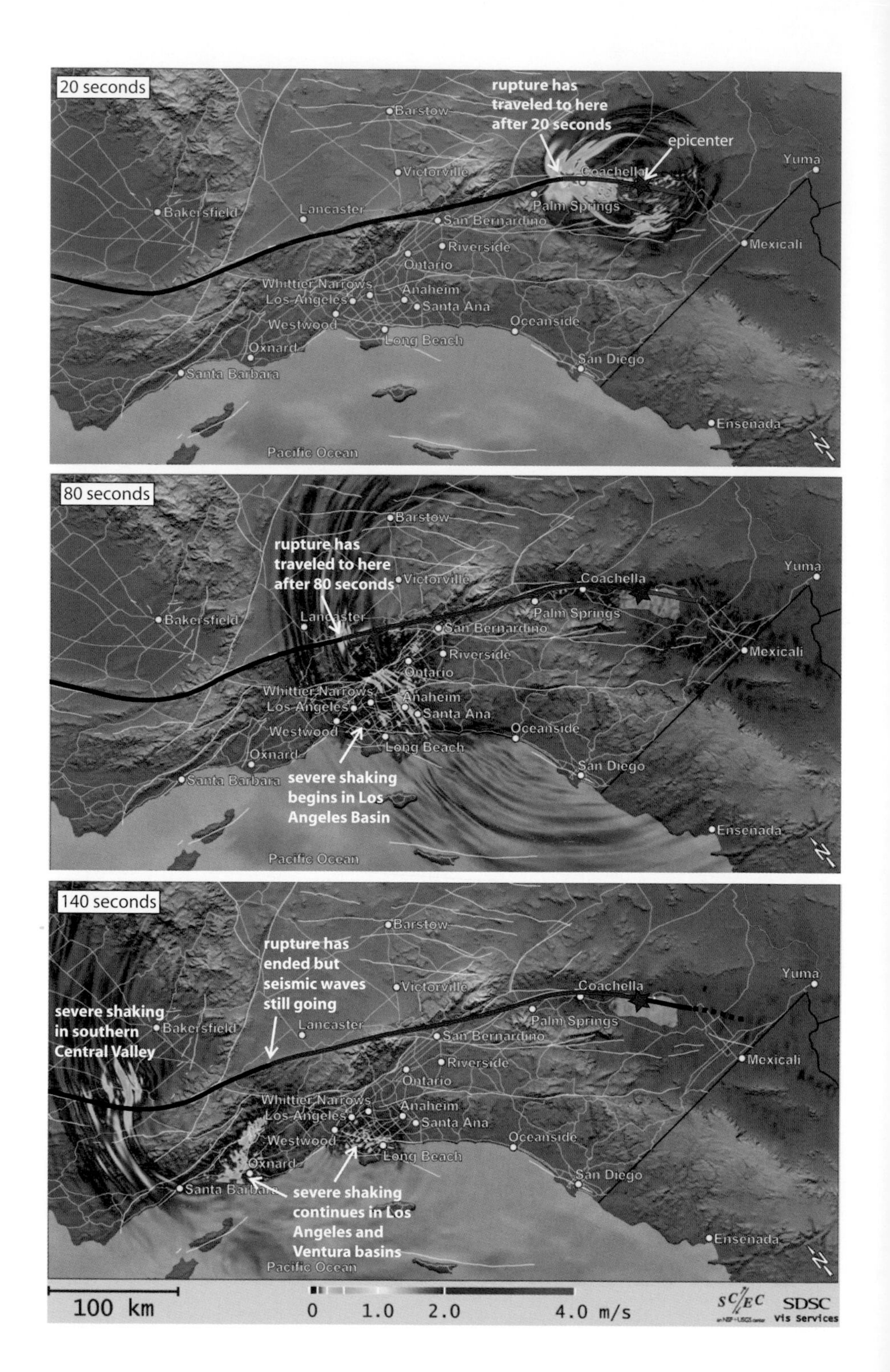

20 seconds
rupture has traveled to here after 20 seconds
epicenter
Barstow
Victorville
Coachella
Palm Springs
Yuma
Bakersfield
Lancaster
San Bernardino
Riverside
Mexicali
Ontario
Whittier Narrows
Los Angeles
Anaheim
Santa Ana
Westwood
Oceanside
Long Beach
Oxnard
San Diego
Santa Barbara
Ensenada
Pacific Ocean
80 seconds
rupture has traveled to here after 80 seconds
severe shaking begins in Los Angeles Basin
140 seconds
rupture has ended but seismic waves still going
severe shaking in southern Central Valley
severe shaking continues in Los Angeles and Ventura basins
100 km
0
1.0
2.0
4.0 m/s
SCEC
SDSC
Vis Services

consequence of the rocks here being crushed within this tectonic vise. Witness some results: During the 1971 San Fernando earthquake (M6.5), the Santa Susana Mountains leapt six feet—and then added another two feet during the Northridge quake in 1994 (M6.7). The coastline near Ventura is squeezing upward faster than anywhere along the California coast—more than one-quarter inch per year, as shown by old marine terraces (notches cut in the land where waves used to break; see chapter 7) that are now well above the sea.

A comprehensive survey of all the earthquake faults that threaten coastal Southern California could form a book of its own. But I can make a few generalizations. First, the main faults to worry about are the *long* ones—faults that stretch at least a few tens of miles—because longer faults can make bigger quakes. While it's not always clear whether a continuous zone of faulted rock represents a single fault (versus several smaller faults closely aligned), the length criterion reduces the pool of worrisome faults to the dozen or so listed in table 3.1. For these faults, we want to know two things: the *maximum likely magnitude* and the *recurrence interval* (the average time between earthquakes). In other words, what is the biggest quake a particular fault is likely to produce, and how often might a quake of that size happen? Neither question is easy to answer. Reliable records of earthquakes in Southern California reach back only two centuries, and although plenty of large quakes have happened during that time (see figure 3.3), the historical record is too brief to give us reliable estimates of either maximum likely magnitude or recurrence interval.

So we rely, instead, on three types of proxy information: fault length, active ground deformation, and slip rate. History and geophysical theory tell us that fault length correlates roughly with maximum magnitude. Ground deformation gives us a handle on how fast strain may

FIGURE 3.5 (OPPOSITE). Computer model of a magnitude 7.8 earthquake initiating on the southernmost San Andreas fault and propagating north. The images show expected ground shaking at 20 seconds (upper), 80 seconds (middle), and 140 seconds (lower) after initiation, with warmer colors representing greater levels of shaking. Notice in the lower image that, although the rupture has stopped and the seismic wave fronts have passed north into the Central Valley, the Los Angeles and Ventura basins are still jiggling like bowls of Jell-O. (Image produced by Amit Chourasia, Kim Olsen, Yifeng Cui, Jing Zhu, David Okaya, Luis Dalguer, Steve Day, Philip Maechling, and Thomas Jordan as part of the ShakeOut project by the Southern California Earthquake Center and the San Diego Supercomputer Center. Labels added by the author.)

TABLE 3.1 SLIP RATES AND EXPECTED MAXIMUM EARTHQUAKE MAGNITUDES FOR SELECTED LONG COASTAL FAULTS IN SOUTHERN CALIFORNIA
Faults listed from north in Santa Barbara County to south in San Diego County

Fault[a]	Approximate length (miles)	Slip rate in millimeters per year (25.4 mm = 1 inch)	Maximum magnitude earthquake that the fault is probably capable of producing
Santa Ynez: western section	39	2.0 ± 1.0	7.1
Santa Ynez: eastern section	41	2.0 ± 1.0	7.1
Mission Ridge–Santa Ana	42	0.4 ± 0.2	7.2
San Cayetano	25	6.0 ± 3.0	7.0
Oak Ridge: offshore section	23	3.0 ± 3.0	7.1
Oak Ridge: onshore section	29	4.0 ± 2.0	7.0
Santa Cruz Island	30	1.0 ± 0.5	7.0
Malibu Coast	22	0.3 ± 0.2	6.7
Santa Monica	17	1.0 ± 0.5	6.6
Palos Verdes	58	3.0 ± 1.0	7.3
Newport–Inglewood: onshore section	40	1.0 ± 0.5	7.1
Newport–Inglewood: offshore section	40	1.5 ± 0.5	7.1
Rose Canyon	42	1.5 ± 0.5	7.2
San Diego Trough: offshore	37	1.5 ± 0.5	7.0
Southernmost San Andreas: Salton Sea to San Bernardino[b]	120	25.0 ± 5.0	7.8

[a] Many of these faults are labeled on the map of Southern California at the front of the book.
[b] Included for comparison. Notice that the slip rate of the southernmost San Andreas is much greater than that of the coastal faults, suggesting that it will produce violent earthquakes more often. But notice too that the expected magnitudes of earthquakes on the coastal faults are projected to be only a little less than magnitudes on the San Andreas.

be accumulating along a fault, just as you can get a sense of when a stick in your hands will break by watching how fast it bends. The slip rate—how rapidly the fault has moved in the past—lets us estimate the recurrence interval. Faster slip rates mean that strain builds more quickly between earthquakes, so quakes are likely to happen more often. Slip rate is the most difficult of the three values to measure because it requires finding datable time-markers that a fault has offset. For example, if a fault has shifted a layer of ancient soil or a stream channel by one thousand inches, and radiocarbon dating (see chapter 7 for an explanation of that) shows that the soil or the channel is one thousand years old, we can say that the fault's slip rate is one inch per year. This doesn't mean that the fault actually moves one inch each year. The slip rate is a long-term average. (By analogy, think about the average speed of your car over its lifetime, taking into account both the short times when it's moving and the much longer times when it sits still.) Most faults sit silent for many years between violent lurches.

Table 3.1 lists the major coastal faults of Southern California with their lengths, estimated slip rates, and estimated maximum magnitudes. I include the southernmost San Andreas fault for comparison. (The table includes only one fault that lies entirely offshore—the San Diego Trough fault—because slip rates of offshore faults are very difficult to determine. To date, the San Diego Trough fault is the only one with a reliable slip rate measurement.) The message from table 3.1 is clear: The estimated slip rates for all of Southern California's coastal faults are far less than for the southernmost San Andreas. This means that the faults shouldn't rupture very often; in fact, thousands of years may pass between large quakes on any given coastal fault. The bad news from table 3.1 is that *all* of the coastal faults listed are capable of unleashing large earthquakes, often M7 or greater, which is comparable to the largest San Andreas quakes. Bottom line: Although the slip rates of Southern California's coastal faults suggest that centuries or millennia can pass between big earthquakes, all of them are capable of doling out a major thrashing.

The Rose Canyon fault (figure 3.6), not far from my home near San Diego, typifies this good news–bad news dichotomy for Southern California's coastal faults (i.e., infrequent but potentially very large earthquakes). A restraining bend where the fault angles west near La Jolla (you can look back at figure 2.4 for an explanation of restraining bends) has pushed La Jolla two miles out to sea and jacked Mount Soledad eight hundred feet above sea level. Notice in figure 3.6 how San Clemente Creek flows nearly straight west until it bumps into Mount Soledad, whereupon it turns sharply south to reach the ocean at Mission Bay. It's an oddly long route; why doesn't San Clemente Creek just continue west to the ocean? The answer is that at one time, it surely did—until the Rose Canyon fault put Mount Soledad in its way. Such a profound rearrangement of local geography probably took many hundreds of large earthquakes along the Rose Canyon fault. (Offset rock units along the fault indicate that the fault has been active for at least four million years.) Yet the Rose Canyon fault has been quiet in historical time. How often do large quakes occur on this fault?

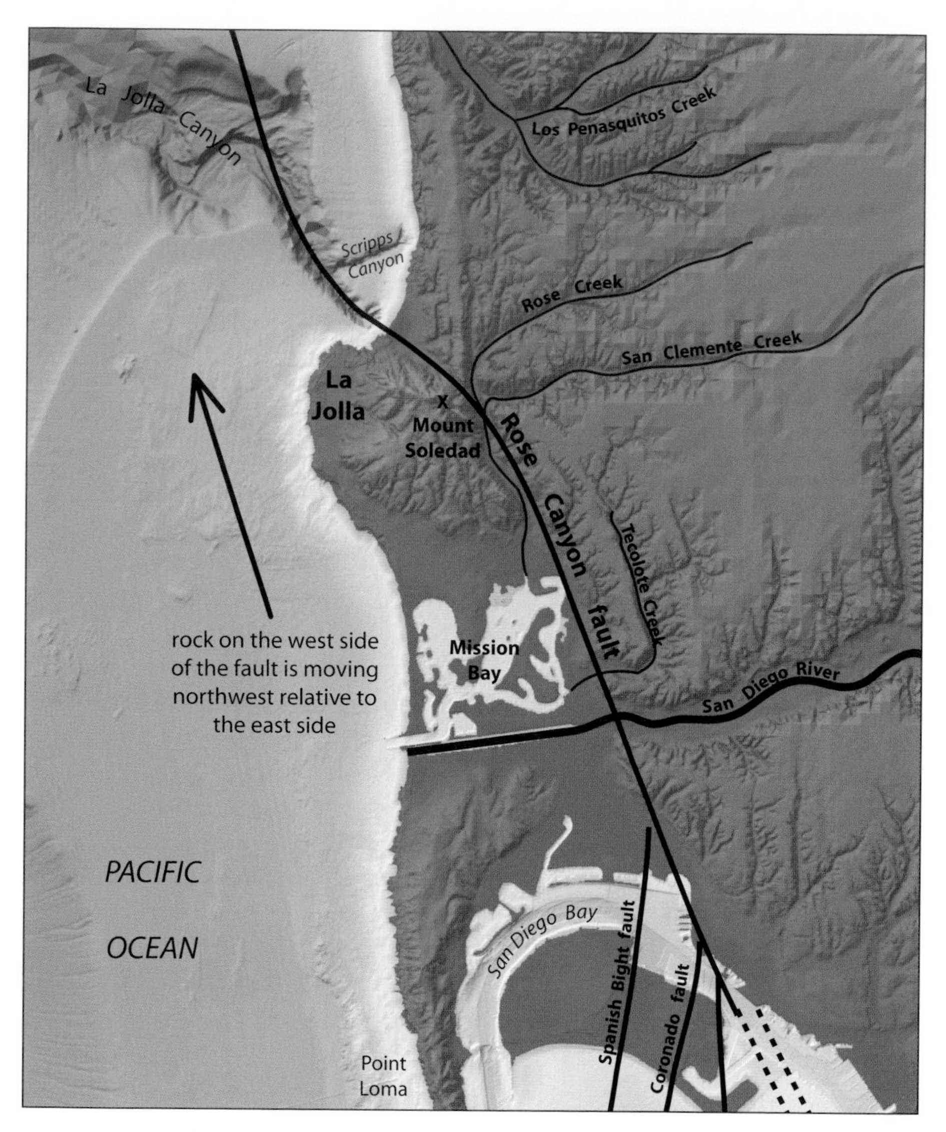

FIGURE 3.6. San Diego's Rose Canyon fault, with its low slip rate but high potential magnitude, typifies many Southern California coastal faults. Compression along a restraining bend (the prominent bend just east of La Jolla) has pushed up Mount Soledad and shoved the land westward to make the stubby headland at La Jolla. San Clemente Creek probably once flowed west straight to the ocean until the rise of Mount Soledad forced it onto its present route south to Mission Bay. In detail, the fault consists of several nearly parallel strands—not a single strand as shown in this simplified view. (Shaded relief image by NOAA, with labels added.)

San Diego geologist Tom Rockwell and colleagues have tackled that question by digging trenches across the Rose Canyon fault to expose old soil beds and stream channels that the fault displaced during prehistoric earthquakes. By figuring out the ages of these displaced features (using radiocarbon dating on fossilized plant remains, for example), Rockwell has concluded that the fault has moved at least twenty-nine feet in the past 8,100 years. This translates to a slip rate of about one to two millimeters per year, which is very low—roughly twenty times lower than the southernmost San Andreas fault (table 3.1). It tells us that the Rose Canyon fault probably doesn't accumulate strain fast enough to fire off a large earthquake more often than about every fifteen hundred to three thousand years. Rockwell has unearthed evidence for a major earthquake sometime between two hundred fifty and five hundred years ago, when the fault may have leapt *ten feet* in a single quake! Were that to happen today, it would decimate large swaths of urban San Diego. But given the Rose Canyon fault's low slip rate—and, thus, slow rate of strain accumulation—San Diegans can probably rest easy (for now).

In this chapter, we explored recent and potential future earthquakes in Southern California. In the next, we'll go back through deep time to investigate the transformations that millions of years of earthquake violence have wrought.

4

Disassembling Southern California

For an extremely large percentage of the history of the world, there was no California. . . . Then, a piece at a time—according to present theory—parts began to assemble. An island arc here, a piece of a continent there—a Japan at a time, a New Zealand, a Madagascar—came crunching in upon the continent and have thus far adhered.

—John McPhee, "Assembling California"
(in *Annals of the Former World*)

California is built largely of imported geology—rock that has traveled here from far away. I told you one such story in chapter 1, when we visited rocks on San Miguel Island that have traveled five hundred miles from Mexico. Five hundred miles may seem like a long way, but the evidence shows that much of California is built of rock that has traveled *thousands* of miles to get here, carried by the Earth's tectonic plates. If you've spent any time roaming around California, you've seen these far-flung immigrants in road-cuts and mountainsides. (I'm talking here about the bedrock that underlies the state, not the veneer of locally derived sediments and lava layers that sometimes cover the bedrock.) The story of how these imported rocks assembled the state is nothing less than the story of California's creation. After we explore that story, we'll zoom in on Southern California to see how a portion of this imported rock was later *disassembled* by the Pacific Plate to create the fault-shattered, earthquake-rattled Southern California of today.

Figure 4.1 shows a fifty-two-year-old California immigrant (originally from Massachusetts) next to a two-hundred-million-year-old

FIGURE 4.1. Here I am embracing two-hundred-million-year-old pillow basalt in the western Sierra Nevada foothills about forty miles northeast of Sacramento. This outcrop, although more than a hundred miles from the nearest deep ocean floor, looks just like the blobby, pillow-shaped lava formations that erupt today at mid-ocean ridges. It originally formed far out in the ancient Pacific Ocean. Subduction carried it to North America's growing western edge during Jurassic time, where it scraped off the deep seabed to join other dispossessed rocks in the process of assembling California. The outcrop is part of the Lake Combie Ophiolite exposed between Clipper Gap and Applegate. (Photograph by Susan Brown.)

California immigrant (originally from somewhere far out in the Pacific Ocean). The rock is pillow basalt, a blobby lava formation that forms wherever molten rock erupts underwater. Pillow basalt paves most of the deep ocean floor, and that makes it the most common type of rock on Earth because the deep ocean floor makes up 60 percent of the planet's surface. The rock in figure 4.1 is unquestionably from the deep ocean floor; both its chemistry and the other deep-sea rocks around it tell us that. Now, though, it's far from home, marooned in mountains that didn't yet exist when it formed on the abyssal seabed back when dinosaurs roamed the Earth. From the coast to the Sierra Nevada, much of California is built of rock that once lay far to the west, on the deep floor of the ancient Pacific. To see how it collected here, we need to look at how the Earth creates and destroys its ocean floor.

SEAFLOOR SPREADING, SUBDUCTION, AND THE ASSEMBLY OF CALIFORNIA

Pillow basalt forms from lava eruptions at mid-ocean ridges—broad undersea mountain belts that wrap around the planet like the seams on a baseball. Deep rift valleys run down the centers of these ridges. The ocean floor slowly splits at these rifts, and lava wells up into the gap to congeal like blood in a wound. The lava squeezes out as toothpaste-like gobs, glowing red for an instant before the cold seawater quenches it into mounds about the size of a bed pillow. Take a submarine down to the central rift valleys of the Mid-Atlantic Ridge, East Pacific Rise, or Juan de Fuca Ridge (shown in figure 4.2a) and you can see the seabed splitting and the lava erupting and forming pillows. Cruise along the deep seabed in either direction away from the central rift valleys, and the pillows become older and more deeply buried in seabed muck. That muck—technically known as *ooze*—is made mostly of microscopic plankton remains. Geologists, by puncturing the deep seabed with thousands of drill holes, have learned two important things about this ooze and the pillow basalt underneath it. First, drill down through enough ooze and you eventually hit pillow basalt. Second, the farther you go from a mid ocean ridge, the thicker the ooze and the older the pillow basalt beneath it.

These facts fit neatly into the concept of *seafloor spreading*—the Earth's system for manufacturing new ocean floor (figure 4.2b). After lava erupts at a mid-ocean ridge to form pillows, it makes room for new lava by spreading away in opposite directions, like two diverging conveyor belts. As the seabed spreads from the ridge, ooze slowly buries the pillows under ever thicker layers. The world's mid-ocean ridges spread in this conveyor belt manner about two inches per year, on average. (The slowest mid-ocean ridges spread about a half-inch per year, while the fastest ridge—the East Pacific Rise—spreads more than six inches per year in some places.) That's a plodding pace in human time, to be sure—but so what? Take two inches per year over the Earth's entire forty-thousand-mile-long mid-ocean ridge system, and watch what happens when geologic time takes over. In 2.4 million years, the Earth will manufacture an area of ocean floor equal to that of the lower forty-eight U.S. states. And in 158 million years—a span less than 4 percent of the age of the Earth—that pace of seafloor spreading will create an area of ocean floor equal to the surface area of the entire planet! (See seafloor spreading text box for more.)

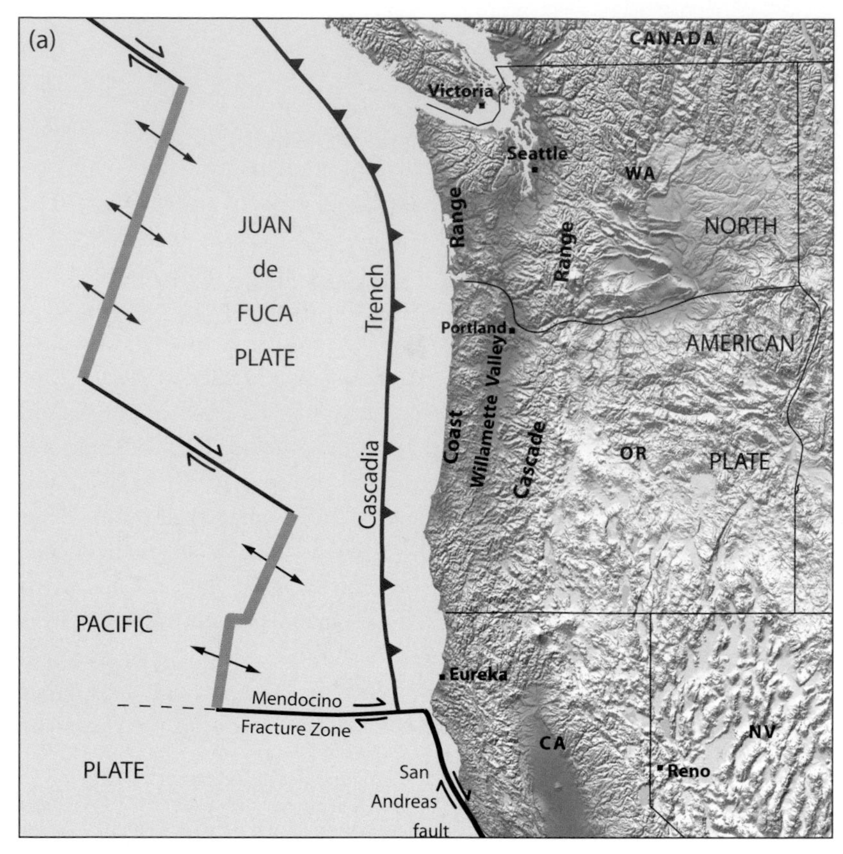

FIGURE 4.2. The Cascadia Subduction Zone off the coast of northernmost California, Oregon, and Washington forms a living example of what was happening along the edge of California during much of Jurassic, Cretaceous, and early Cenozoic time—a span from about 160 million to 50 million years ago.
(a) The Cascadia Subduction Zone extends from Cape Mendocino in northern California north to Vancouver Island. In this system, the oceanic Juan de Fuca Plate grows by seafloor spreading at the Juan de Fuca Ridge and then subducts under the North American Plate at the Cascadia Trench. Notice the locations of the Coast Range, Willamette Valley, and Cascade Range, whose roles in the system are shown in diagram b. (Shaded relief image from NASA, with labels added.)

What happens to all of this newly minted ocean floor? With the seabed constantly splitting and growing from its mid-ocean ridges, you might expect the planet to be swelling like a balloon. But it's not, because the ocean floor is destroyed apace with its creation. This happens at ocean trenches—great troughs where the ocean floor bends down beneath a neighboring plate to plunge into the Earth's hot

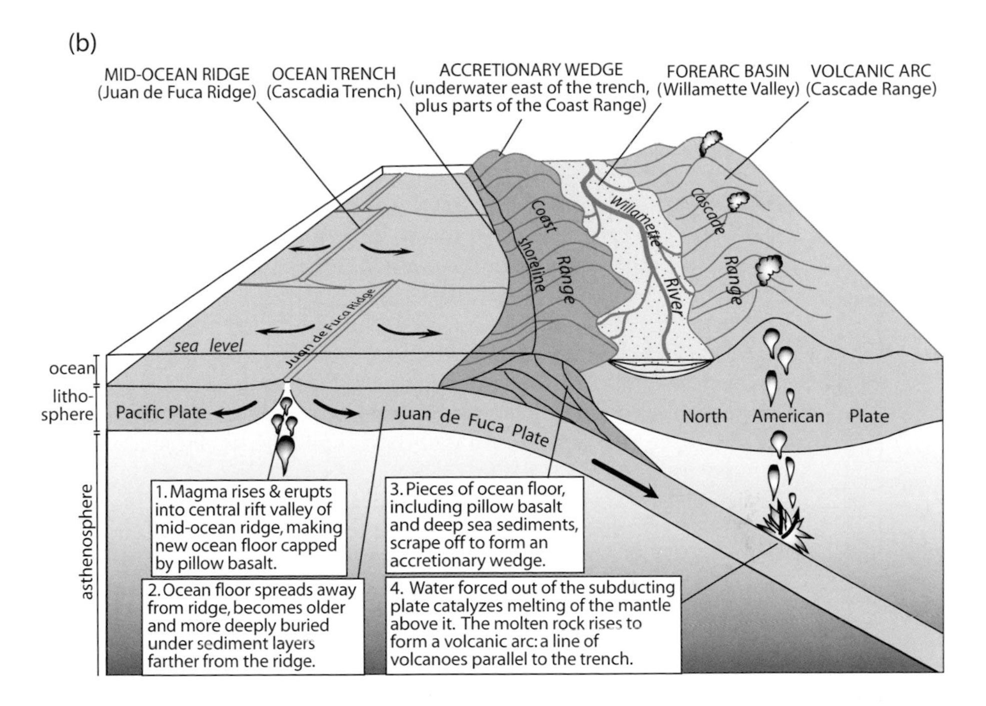

FIGURE 4.2 *(continued).* (b) The components of the Cascadia Subduction Zone portrayed along an east-west slice into the Earth. Where the Juan de Fuca Plate dives into the Cascadia Trench, seabed sediments and pieces of oceanic crust, including pillow basalt, scrape off to form a growing accretionary wedge that rises out of the sea to form part of the Coast Range. The Willamette Valley is a fore-arc basin: a trough that receives sediments shed from the flanking ranges. The Cascade Range is a volcanic arc fed by magma rising from above the subducting Juan de Fuca Plate. Fossil examples of all three components—accretionary wedge, fore-arc basin, and the deeply eroded roots of a volcanic arc—make up the bedrock geology of much of California today.

interior (figure 4.2b). This process—called *subduction*—consumes the ocean floor almost as fast as seafloor spreading produces it.

Almost.

Subduction, it turns out, doesn't eat up every square foot of ocean floor churned out by seafloor spreading. There is a small—but crucial—imbalance between the creation of the ocean floor and its destruction. That imbalance, among other things, has created much of westernmost North America, including California.

As the seafloor slides into an ocean trench, the edge of the adjacent plate can act like a plow blade, scraping off slivers of seafloor rock. Anything sticking up from the seafloor—submarine plateaus, seamounts,

Seafloor Spreading: Doing the Math

Take the global average seafloor spreading rate of two inches per year, and multiply that by the forty-thousand-mile length of the mid-ocean ridge system, and you get a global production rate of 1.25 square miles of new ocean floor per year. Let's assume that this is a typical rate over geologic time (although spreading rates were likely a bit faster when the Earth was younger and hotter). We can use this value to calculate how long it would take the Earth to manufacture certain amounts of new ocean floor. For instance, the area of the lower forty-eight U.S. states is about three million square miles; the time it would take seafloor spreading to make this amount of ocean floor is calculated as 3×10^6 square miles ÷ 1.25 square miles per year = 2.4 million years. The surface area of the Earth is about 197 million square miles; seafloor spreading would make this amount of ocean floor in 197×10^6 square miles ÷ 1.25 square miles per year = 158 million years. The Earth is about 4.56 billion years old, so we can surmise that a truly stupendous amount of ocean floor—an amount many times greater than the surface area of the Earth—has been created and destroyed over geologic time.

or even entire islands—can get scraped off as well. The astonishing result—played out over the scope of geologic time—is that *continents grow bigger as pieces of oceanic rock collect against their edges*. Bodies of imported rock that moving tectonic plates strand along the edges of continents are known as *accreted terranes*. Stand anywhere in western Mexico, California, Oregon, Washington, British Columbia, or Alaska, and you stand—more often than not—on accreted terranes: bodies of rock that once lay far out in the Pacific Ocean that have been marooned against North America's western edge (figures 4.3 and 4.4).

The reason this has happened comes down to North America's long history of westward migration, combined with the eastward migration of pieces of old ocean floor that have collided with the edge of the continent or slid underneath it. Two hundred and fifty million years ago, our continent was stuck firmly to Eurasia and Africa as part of the supercontinent Pangaea. But about two hundred million years ago, North America began to tear away and head west, opening the Atlantic Ocean in its wake. New ocean floor created by seafloor spreading in the growing Atlantic had to be balanced by destruction of ocean floor in the Pacific. That happened as the continent overrode multiple pieces of ancient

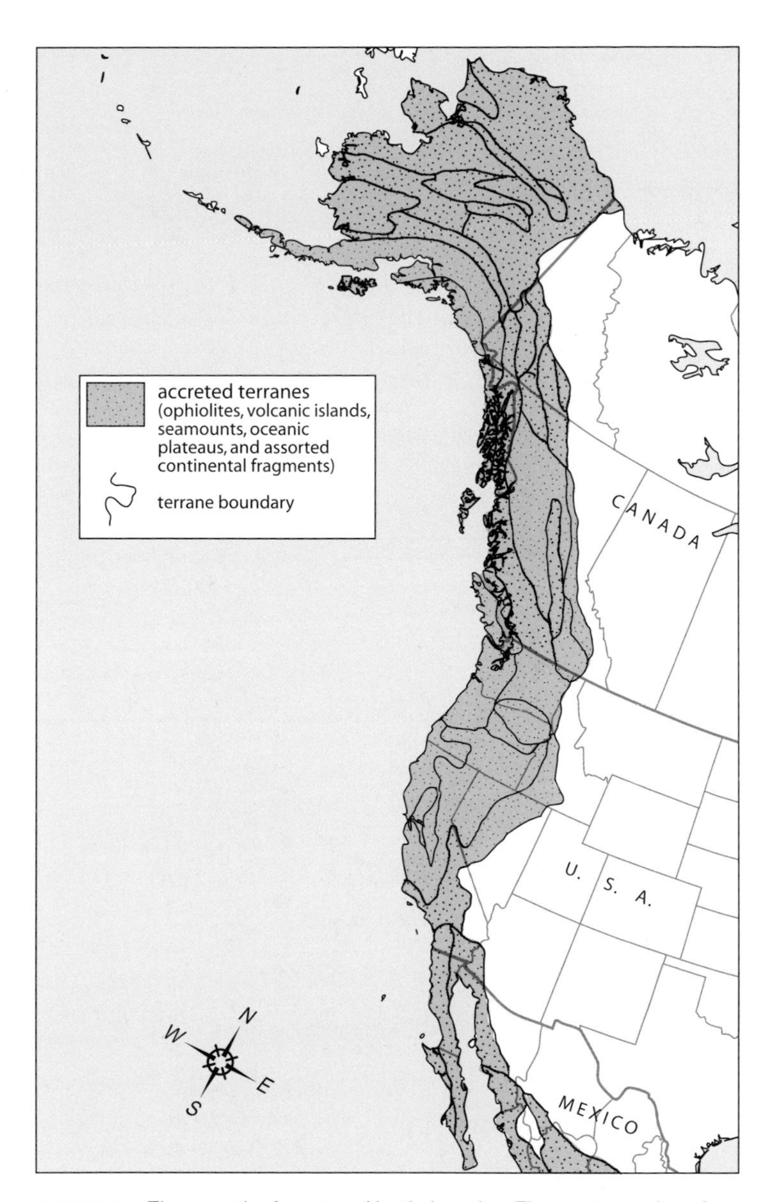

FIGURE 4.3. The growth of western North America. The western edge of the continent comprises vast strips of immigrant rock, called *accreted terranes,* brought here by subduction from far out in the ancient Pacific Ocean. Before they docked onto the continent, these terranes were pieces of oceanic crust, volcanic islands, seamounts, oceanic plateaus, and fragments of small continents. Like groceries piling up at the end of a checkout-line conveyor belt, they have collected, one behind the other, against the continent's western edge during the past two hundred million years. We distinguish individual accreted terranes by the great faults that separate them and by the far-flung rocks and fossils within them. (From Meldahl 2011.)

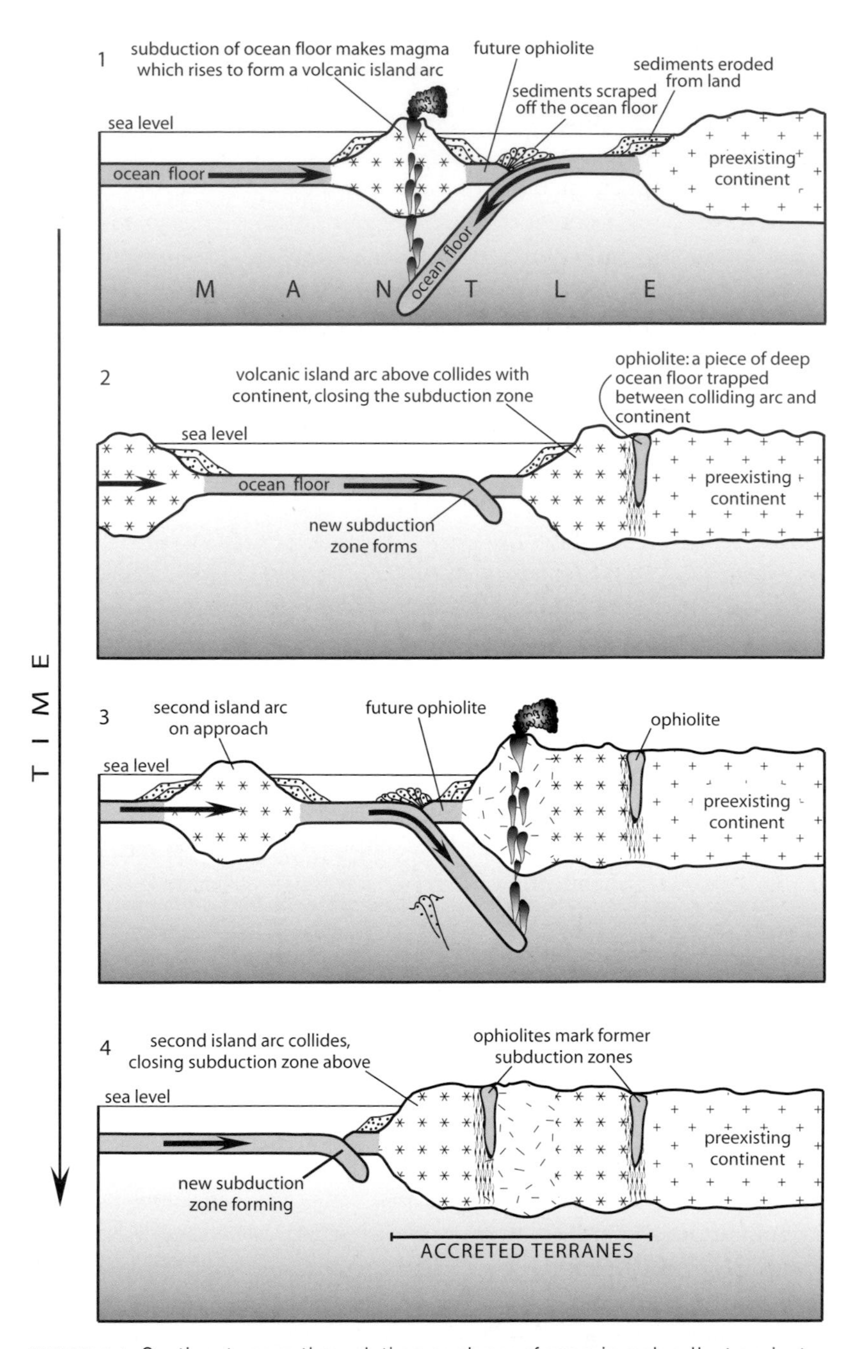

FIGURE 4.4. Continents grow through time as pieces of oceanic rock collect against their edges. This figure illustrates (very schematically) the growth of California and, by extension, much of the rest of western North America. Over the past two hundred million years, subduction has stacked all manner of oceanic real estate against the continent's edge, as shown in map view in figure 4.3. (From Meldahl 2011.)

Pacific Ocean seafloor. Much of this old seabed went under the continent, but several million cubic miles of it glommed onto the continent's western edge to add the new geologic real estate shown in figure 4.3. By the metric of human time, this continental growth happened very slowly. But it didn't happen quietly. The Earth's plates don't glide past each other like well-oiled machine parts; they stick at their edges for a time and then shift violently during earthquakes. (That's why a world map of earthquakes translates to a map of the Earth tectonic plates; see figure 3.1.) California grew thanks to the violent work of numberless earthquakes. As John McPhee once put it, "Earthquakes brought things from far parts of the world to fashion California."

By about one hundred million years ago (the middle of Cretaceous time), California's assembly via earthquakes and subduction was mostly complete. Rocks from this period give us some of our clearest evidence for how subduction built California. Let me take you on a tour of some of these rocks using one of geology's guiding principles. Geologists are always happier about explanations for things in the *past* if we can see the Earth doing similar things *today*—a principle known as *uniformitarianism.* (A simple example: When we see that lava pouring out of Kilauea Volcano chills into basalt rock, it makes us confident that ancient basalt likewise came from chilled lava.) Following this principle, we can fathom much of California's past by looking at what is happening today along the coast of Oregon and Washington. I'll give you the punch line up front: We think that much of the bedrock of California represents a *fossil subduction zone,* much like the active subduction zone we see today along the coast of Oregon and Washington (figure 4.2).

Our geologic tour has six quick stops. The first three are in the active Cascadia Subduction Zone of Oregon. The last three are in Cretaceous rocks of a fossil subduction zone that makes up much of California.

Stop 1: Oregon's Coast Range. Visualize standing on a foggy Oregon coast, where Pacific swells are breaking just a few yards from a dripping forest that rises above the beach four thousand feet to the crest of the Coast Range. Oceanic storms dump more than a hundred inches of rain annually on the Coast Range, so the rocks here are always wet. A few swings of a rock hammer show us that they were *born* wet. Much of the Coast Range is made of pillow basalt and deep-sea sediments that once lay more than a mile below sea level on the Juan de Fuca Plate (figure 4.2a). The pillow basalt originated as lava erupting at the Juan de Fuca Ridge. It then slid east, conveyor-belt-style, as part of the Juan de Fuca Plate, collecting deep-sea sediments (mostly mud and ooze) as it traveled east before diving into the Cascadia Trench. There, the edge of North America scraped the pillows and sediments off the subducting seabed, sort of like a grader scraping dirt off a roadbed. As more scraped-off material wedged in behind, the pillows and deep-sea sediments were gradually levered out of the ocean, so that we find them high above the sea today in the Coast Range. In technical terminology, the Coast Range is an *accretionary wedge:* a fault-riddled jumble of deep-sea rock scraped off a subducting plate and added to the plate next door (see figure 4.2b). Oregon's accretionary wedge represents the slow but ceaseless geologic transfer of pieces of the Juan de Fuca Plate onto North America's still growing edge.

Stop 2: Oregon's Willamette Valley. Heading east over the Coast Range from stop 1, we descend into the Willamette Valley, a low-lying trough that gathers sand and gravel sloughed off the Coast Range, and off the Cascade Range to the east. The Willamette River carries much of this debris north to the Columbia River and onward to the ocean. But much of it stays behind, piling up as the valley sinks to accommodate it. The Willamette Valley represents the region of a subduction zone called the *fore-arc basin* (figure 4.2b)—so named because it lies forward of the volcanic arc, represented by the volcanoes of the Cascade Range. Fore-arc basins may be above sea level, like the Willamette Valley, or below it, but either way, they act as receptacles for sediments shed from the flanking accretionary wedge and volcanic arc.

Stop 3: Oregon's Cascade Range. Crossing the Willamette Valley, we climb the volcanic slopes of the Cascade Range. As the Juan de Fuca Plate descends like a down-escalator beneath Washington, Oregon, and northernmost California, the rising heat and pressure drive seawater out of the upper part of the plate into mantle above (figure 4.2b). (The seawater is trapped in seabed sediments, in fractures, and in hydrous [water-bearing] minerals in the down-going plate.) The Earth's mantle, despite being very hot, is normally solid; we know this because certain types of seismic waves that can't pass through fluids travel readily through the mantle. But if you inject seawater into hot mantle rock under intense pressure, a curious thing happens: Some of the rock melts. This is because water molecules under high pressure jam into the molecular spaces of minerals within mantle rock, interfering with the chemical bonds that hold the minerals together, thus causing some of the rock to melt. This fresh-made magma, spawned by subducted seawater, rises buoyantly from its origin about sixty miles underground and slowly punches its way upward into the crust above. (Magma is always less dense than the rock from which it comes, so where it forms, it rises.) Some of the magma squirts all the way to the surface, where it erupts periodically to form the volcanic cones of the Cascade Range—Mount Shasta, the Three Sisters, Mount Hood, Mount St. Helens, Mount Rainier, and others. But most of the magma doesn't make it to the surface. Instead, it solidifies underground into granite and related rocks that form a foundation for the volcanoes above. Cut five or more miles of volcanic rock off the Cascade Range, and you will see granite, along with magma chilling into yet more granite.

Now we head south to California, where we will find *fossil* versions of the same three subduction zone components (accretionary wedge, fore-arc basin, volcanic arc) that are actively forming in Oregon today. The rocks range from about 150 million to 50 million years old, and the subduction that made them has long since ceased. But the similarities to Oregon will give you déjà vu. Figure 4.5 shows you a map of these rocks.

Stop 4: California Coast near San Francisco. Here, in a rock formation called the Franciscan Complex, we find pillow basalts and deep-sea sediments that look a lot like those in Oregon's Coast Range (stop 1). The Franciscan Complex, in other words, looks like an ancient accretionary wedge. Before the discovery of plate tectonics during the 1960s, the Franciscan Complex was a geologic conundrum—a mixed-up, fault-shattered wreckage of rock whose origin defied explanation. But in the late 1960s, a young Stanford geologist named William Dickinson, using the fresh ideas of plate tectonic theory, showed that the Franciscan Complex is what you would expect to find where subduction (a process that had just been discovered) takes quadrillions of tons of deep ocean floor and scrapes it off against the edge of a continent. Our understanding of California—and, indeed, of the geologic formation of the entire western United States—took a gigantic leap forward at that moment, because it meant that subduction had once been active along almost the entire western edge of North America. As Dickinson later wrote, "It was as if we had spent decades slowly climbing a high hill to attain a vantage point, and with a sudden burst of intellectual energy reached the summit from which the whole landscape was laid out before us."

Stop 5: Great Central Valley. Heading east from the Franciscan Complex at stop 4, we find that it wedges beneath a different rock formation

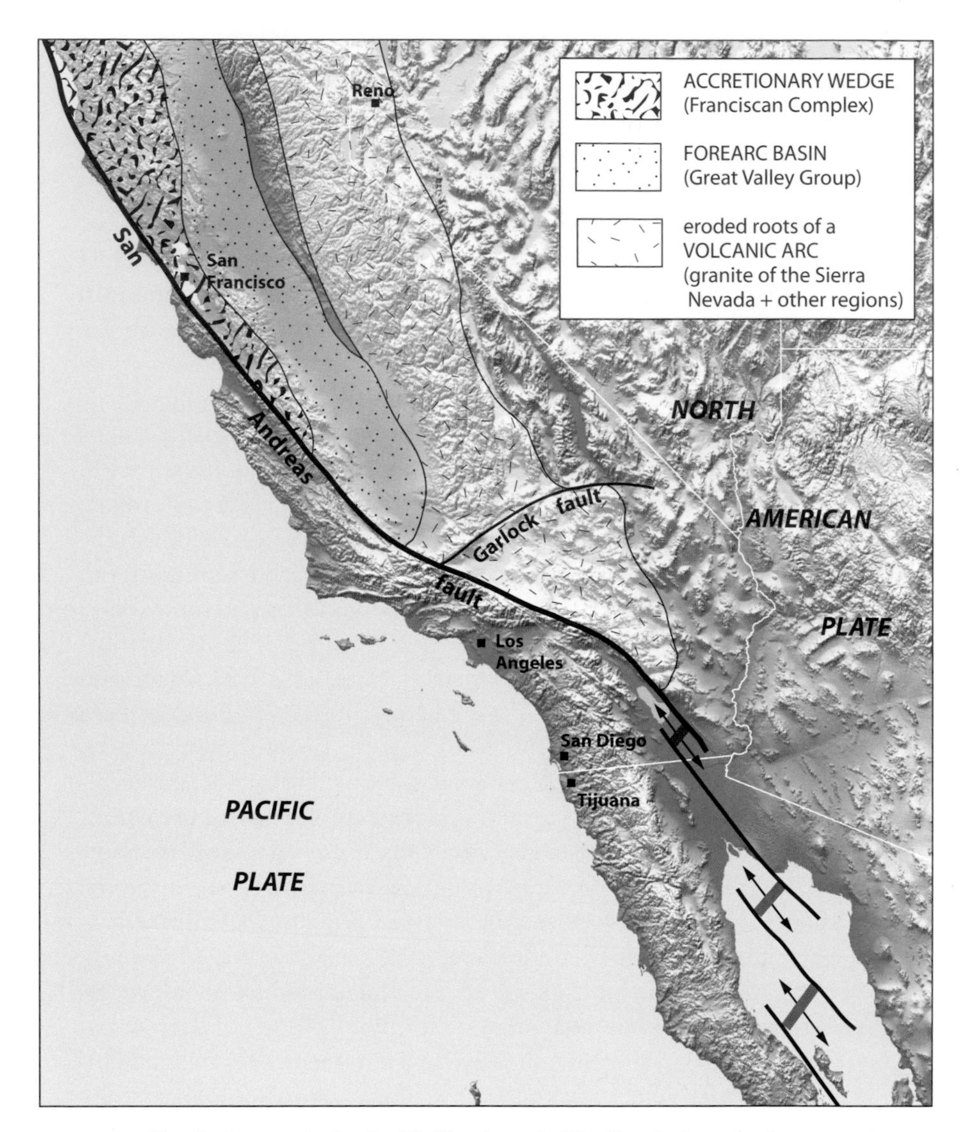

FIGURE 4.5. The Cretaceous bedrock of California east of the San Andreas fault represents an intact fossil subduction zone. From west to east, we recognize the remains of an accretionary wedge (the Franciscan Complex), a fore-arc basin (the Great Valley Group), and the deeply eroded roots of a volcanic arc (the granite of the Sierra Nevada). These same three belts also exist west of the San Andreas fault, but there they have been dismembered as they were captured by the Pacific Plate, as shown in figure 4.7. (Shaded relief image from NASA, with labels added; based on Dickinson 2003.)

called the Great Valley Group, named for prominent up-tilted exposures along the western edge of the Great Central Valley. Here we find stunning outcrops of turbidites: layers of sand and mud that cascaded into the Cretaceous ocean like undersea avalanches. The Great Valley Group looks very much like the sediment layers you would expect to see piling up in Oregon's Willamette Valley if you dropped the valley a bit below sea level. In other words, the Great Valley Group looks like it accumulated in a fossil fore-arc basin—but one that, like the Franciscan Complex, has lost its subduction zone (figure 4.5).

Stop 6: Sierra Nevada. Heading east across the Central Valley, we now climb the granite slopes of the high Sierra Nevada. About one hundred million years ago, Cascade-type volcanoes, spawned by subduction, towered where the Sierra Nevada range is today. Erosion has long since chewed those volcanoes into tiny pieces and sent them to the Pacific Ocean (although remnant volcanic layers remain in some parts of the Sierra). The granite that once lay below those volcanoes was heaved skyward by mighty faults during the past five to ten million years, so that it now forms the backbone of the mountain range. In other words, the Sierra Nevada range represents the deeply eroded, uplifted roots of an extinct volcanic arc (figure 4.5).

To sum up, the bedrock of large swaths of California comprises a trifold grouping—accretionary wedge, fore-arc basin, and volcanic arc—that bears witness to a now vanished subduction zone. So where did that subduction zone go? The answer is that the San Andreas fault replaced it. That process—the change from millions of years of subduction to more recent side-by-side sliding along the San Andreas fault—took the geologic story of California in a new direction.

THE FARALLON PLATE AND THE SAN ANDREAS FAULT

Perhaps the best way to see how this change occurred is to realize that the Juan de Fuca Plate (figure 4.2) is actually a surviving snippet of a larger plate, called the Farallon Plate, that once subducted continuously along North America's western edge all the way from Canada to Mexico. The subduction and gradual annihilation of the Farallon Plate has been the single most important driver of geologic events in the American West during the past one-hundred-plus million years. It pushed up the Rocky Mountains, stretched the continent to form the Basin and Range, seeded the crust with metal ores, and produced the San Andreas

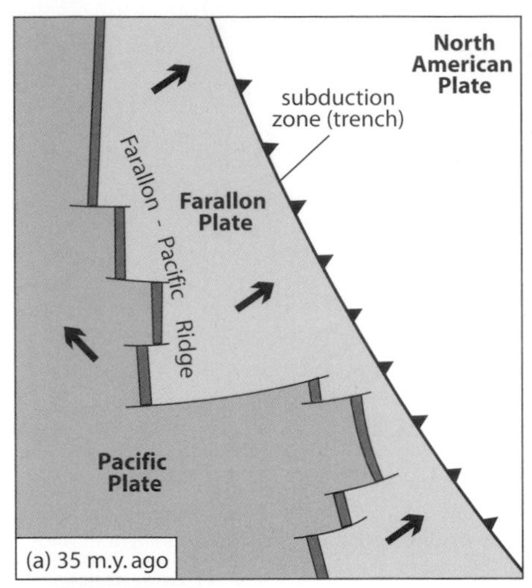

North American Plate
subduction zone (trench)
Farallon - Pacific Ridge
Farallon Plate
Pacific Plate
(a) 35 m.y. ago

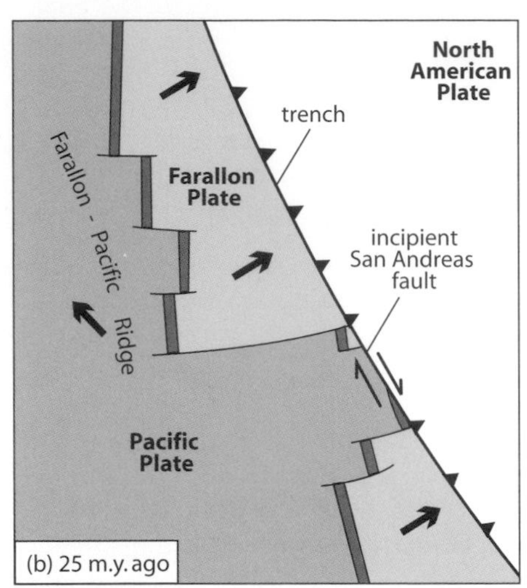

North American Plate
trench
Farallon - Pacific Ridge
Farallon Plate
incipient San Andreas fault
Pacific Plate
(b) 25 m.y. ago

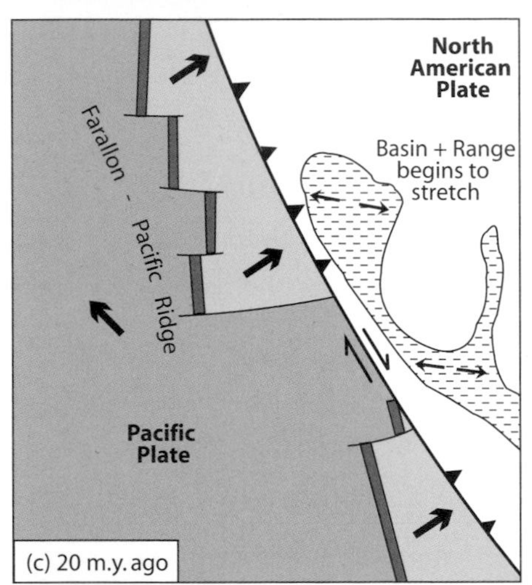

North American Plate
Farallon - Pacific Ridge
Basin + Range begins to stretch
Pacific Plate
(c) 20 m.y. ago

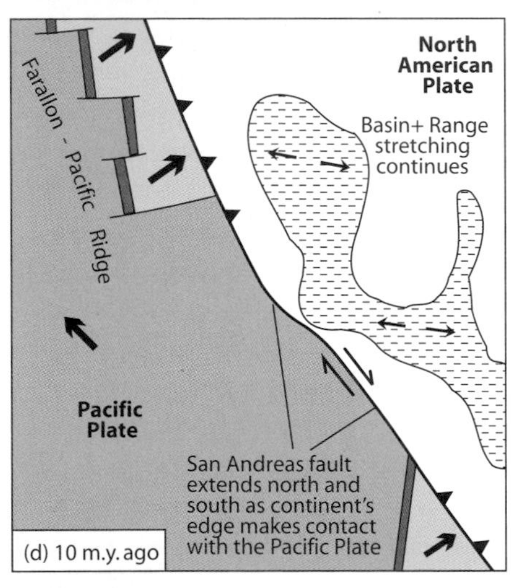

North American Plate
Farallon - Pacific Ridge
Basin+ Range stretching continues
Pacific Plate
San Andreas fault extends north and south as continent's edge makes contact with the Pacific Plate
(d) 10 m.y. ago

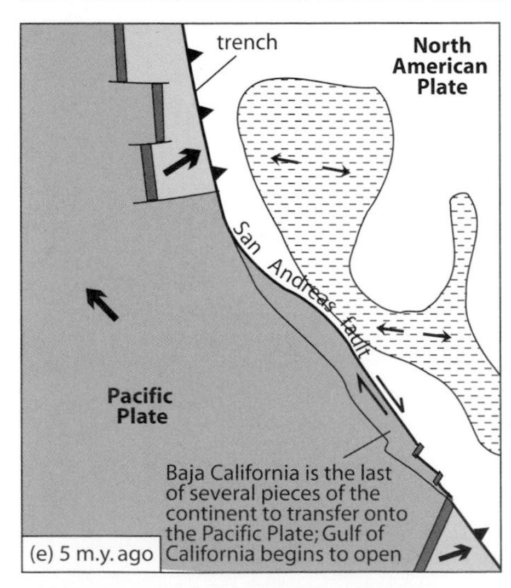

trench
North American Plate
San Andreas fault
Pacific Plate
Baja California is the last of several pieces of the continent to transfer onto the Pacific Plate; Gulf of California begins to open
(e) 5 m.y. ago

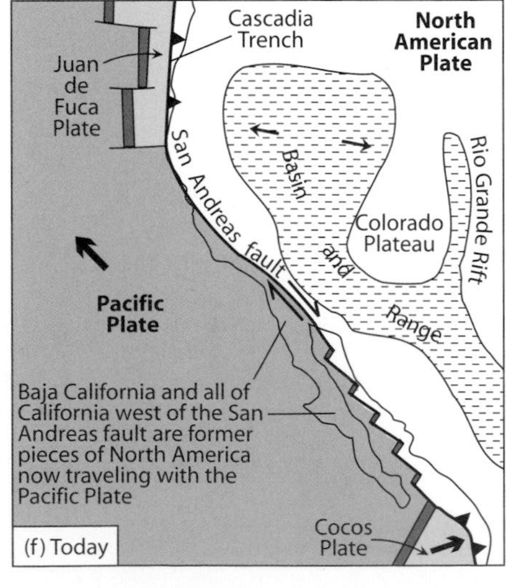

Cascadia Trench
North American Plate
Juan de Fuca Plate
San Andreas fault
Basin and Range
Colorado Plateau
Rio Grande Rift
Pacific Plate
Baja California and all of California west of the San Andreas fault are former pieces of North America now traveling with the Pacific Plate
Cocos Plate
(f) Today

fault. (My previous book *Rough-Hewn Land* explores these stories in detail, if you're interested.) Our focus here is to see how the subduction zone of the Farallon Plate was eliminated to give birth to the San Andreas fault. To see how that happened, look at figure 4.6.

The first panel of figure 4.6 shows the situation that prevailed from Cretaceous time up until about twenty-eight million years ago (Oligocene time). During that span, a continuous mid-ocean ridge—the Farallon–Pacific Ridge—separated the Pacific Plate from the Farallon Plate. The Farallon and Pacific plates grew and spread away from one another at this ridge, with the Farallon Plate moving relatively northeast and the Pacific Plate northwest (arrows in figure 4.6). As the Farallon Plate encountered the western edge of North America, it subducted down an oceanic trench that once ran continuously along the continent's west coast, producing (among other things) the rock formations we saw at stops 4, 5, and 6. But the Farallon Plate was caught in a closing trap. As the North American Plate migrated relatively westward to accommodate the widening Atlantic Ocean, the Farallon Plate was consumed down the trench faster than it grew from the Farallon–Pacific Ridge. Figure 4.6 (panels b–f) shows what happened as a result. Wherever North America's leading edge closed like a sliding trap door over the Farallon Plate and made contact with the Pacific Plate on the other side, the Farallon Plate was eliminated, and the plate boundary changed from subduction to side-by-side sliding. Why, you may wonder, didn't the Pacific Plate just subduct under North America right behind the Farallon Plate? The answer is that its motion in relation to North America was different. Whereas the Farallon and North American plates had been heading toward each other (so that the former dove beneath the

---

FIGURE 4.6 (OPPOSITE). A map-view explanation of how the demise of the Farallon Plate produced the San Andreas fault and led to the capture of several pieces of North America by the Pacific Plate. Before about thirty million years ago, a continuous mid-ocean ridge—the Farallon–Pacific Ridge—separated the Pacific Plate from the Farallon Plate. From this ridge, the Farallon Plate spread eastward before subducting under the North American Plate. Starting about twenty-eight million years ago, the North American Plate began to intersect the Farallon–Pacific Ridge and make contact with the Pacific Plate on the other side. Wherever the two plates made contact, their relative motions resulted not in subduction but in side-by-side sliding at their shared edges. A new type of plate boundary was thus born—a side-by-side sliding boundary that grew, north and south, to become the San Andreas fault and its southern extension, the Gulf of California. As this side-by-side movement developed, several continental pieces that were formerly part of the North American Plate shifted over onto the Pacific Plate. Today, those pieces include Baja California, all of California west of the San Andreas fault, and the Continental Borderland (see figure 1.7). This change in the plate boundary also stretched the continent to form the Basin and Range Province. (From Meldahl 2011; based on Atwater 1970.)

latter), the Pacific and North American plates were going in nearly the same direction. The slight difference in their relative motions—west for the North American Plate, northwest for the Pacific Plate—resulted not in subduction at their touching edges, but in side-by-side sliding. A new plate boundary was thus born, marked today by the San Andreas fault and its southward extension, the Gulf of California. The key thing to understand is that, wherever North America overtopped the Farallon–Pacific Ridge, subduction ended, and side-by-side sliding replaced it. That region of terminated subduction—now occupied by the San Andreas fault and the Gulf of California—presently extends from Cape Mendocino in Northern California south to Puerto Vallarta, Mexico, at the mouth of the Gulf of California.

Two tail-end segments of the Farallon Plate remain in view on the ocean floor today. These are the Juan de Fuca Plate, which plunges down the Cascadia Subduction Zone from Vancouver Island to Cape Mendocino, and the larger Cocos Plate, which subducts under Mexico south of the Gulf of California. (You can see them both in the last panel of figure 4.6, and on the Tectonic Plates map at the front of the book.) Once these shrinking remnants of the Farallon Plate are swallowed by their respective trenches, the Farallon Plate will be no more.

Perhaps now you can see why I used the Cascadia Subduction Zone of Oregon to build a picture of what once happened throughout most of California. The Cascadia Subduction Zone is really just a surviving part of the once much longer subduction zone of the Farallon Plate—one that once extended, unbroken, from Canada to Mexico. The changes portrayed in figure 4.6 eliminated that subduction zone from Cape Mendocino south to Puerto Vallarta, Mexico. But where it survives, it gives us a window into California's past.

THE PACIFIC PLATE KIDNAPS SOUTHERN CALIFORNIA

Now it's time to bring our focus back to Southern California. The story that I want to show you now is how the change from subduction to side-by-side sliding *disassembled the fossil subduction zone to produce the fault-splintered Southern California geology that we see today.* That may seem straightforward, but, as we'll see, the details get a bit involved.

To begin, look back for a moment at figure 4.5, which shows the geology *east* of the San Andreas fault. Here, as we saw at stops 4, 5, and 6 on the tour above, much of the bedrock represents an intact fossil subduction zone that formed above the Farallon Plate long before the San Andreas fault developed. Now shift to figure 4.7, which shows the bedrock *west* of the San Andreas fault. Here we find the same subduction-generated rock units—but with a striking difference. West of the fault, the Pacific Plate has torn asunder the old subduction zone rocks and dragged them several hundred miles northwest. Take those Humpty Dumpty rocks in figure 4.7 and put them back together again, and you'll have reversed more than fifteen million years of tectonic kidnapping by the Pacific Plate.

Figure 4.8 sums up how the geology shown in figure 4.7 came to be. It shows how the Pacific Plate serially captured several pieces of North America as the San Andreas fault evolved. Let me take you through the evidence for two striking developments portrayed in figure 4.8. One development is the dramatic clockwise spin of that big block of crust labeled "WTRB" for the Western Transverse Ranges Block (also shown in figure 4.7). The WTRB is the bedrock that makes up the Santa Ynez and Santa Monica mountains, the Santa Barbara Channel, and the Northern Channel Islands. (If you live in Malibu, Oxnard, Ventura, Santa Barbara, or elsewhere in that part of California, you live on the WTRB.) The other development is the stretching of the crust that occurred in the wake of the WTRB's northwestward migration. This is the area labeled "stretched zone" in figures 4.7 and 4.8, and it makes up the eastern half of the Continental Borderland. For the rest of this chapter, I want to show you the evidence that speaks to these two developments: namely, the spin and northwestward migration of the WTRB and the stretching of the eastern Continental Borderland. If you accept this evidence, then the whole story of Southern California's geologic disassembly by the Pacific Plate falls into place.

First, the clockwise spin and northwestward migration of the WTRB. As figure 4.8 shows, we think the WTRB once lay west of San Diego. That's because, when we rotate the WTRB back into that position, we find that many of the rocks match up, jigsaw-puzzle style, with rocks in San Diego. Moreover, the flow direction of ancient rivers matches up too. Recall from chapter 1 the story of the Eocene riverbed pebbles that migrated from Mexico to San Diego and to the Northern Channel Islands (see figures 1.1 and 1.2). In the WTRB today, these riverbed rocks are oriented such that the rivers appear to have flowed from south to north *away* from the ocean in that part of California, which doesn't make sense. But when we spin the WTRB counterclockwise and put it back to its presumed starting position west of San Diego, the riverbed

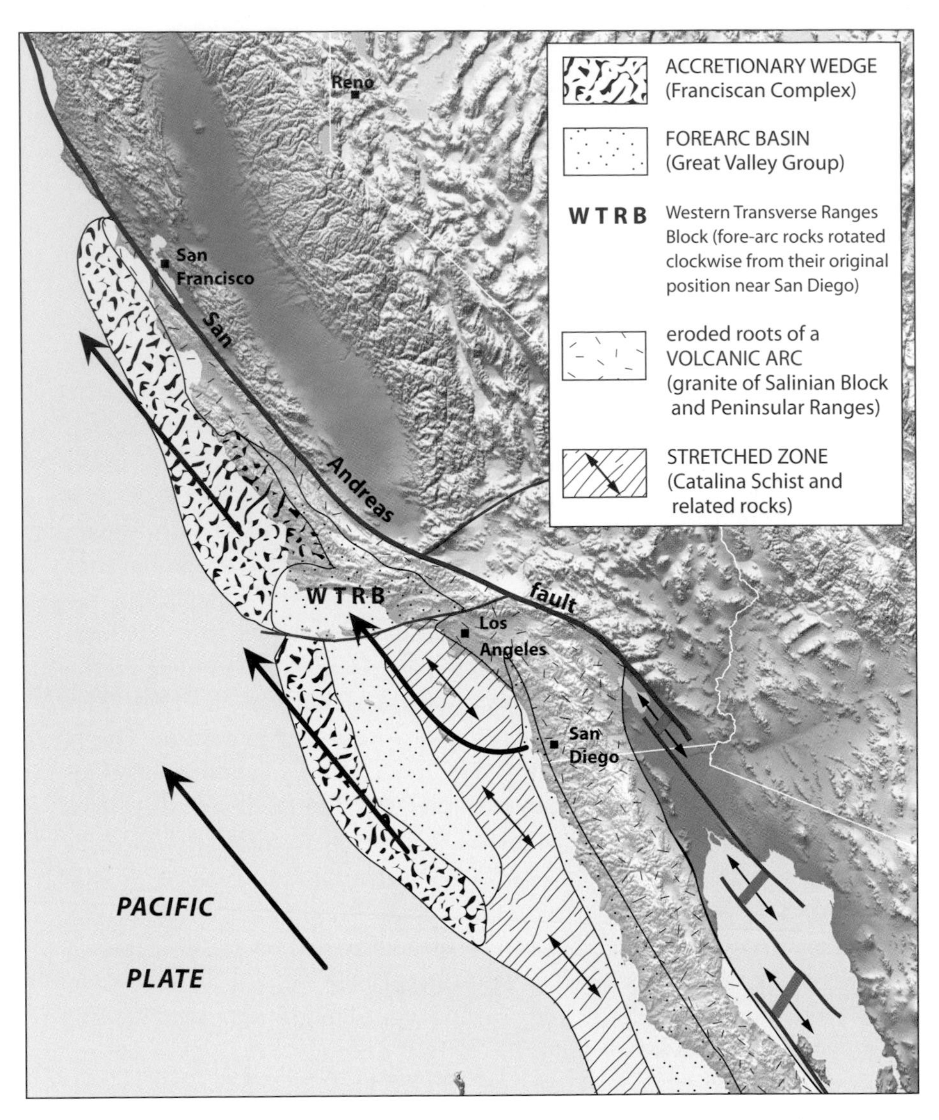

FIGURE 4.7. In Cretaceous-age bedrock west of the San Andreas fault, we find the same subduction zone–related belts that we see to the east of the fault shown in figure 4.5, but with a striking difference. In the process of shifting over to the Pacific Plate during the past twenty million years, the rocks west of the fault have been broken apart and dragged northwest several hundred miles with the Pacific Plate. Two dramatic developments accompanied this change. One was the clockwise rotation of the WTRB (Western Transverse Ranges Block) as it traveled from near San Diego to its present location. The other was widespread stretching of the crust (the area of the map labeled "stretched zone") that occurred in the wake of the WTRB's northwestward migration. For the time sequence that led to this present-day geology, see figure 4.8. (Shaded relief image from NASA, with labels added; based on Fisher et al. 2009a: fig. 4.)

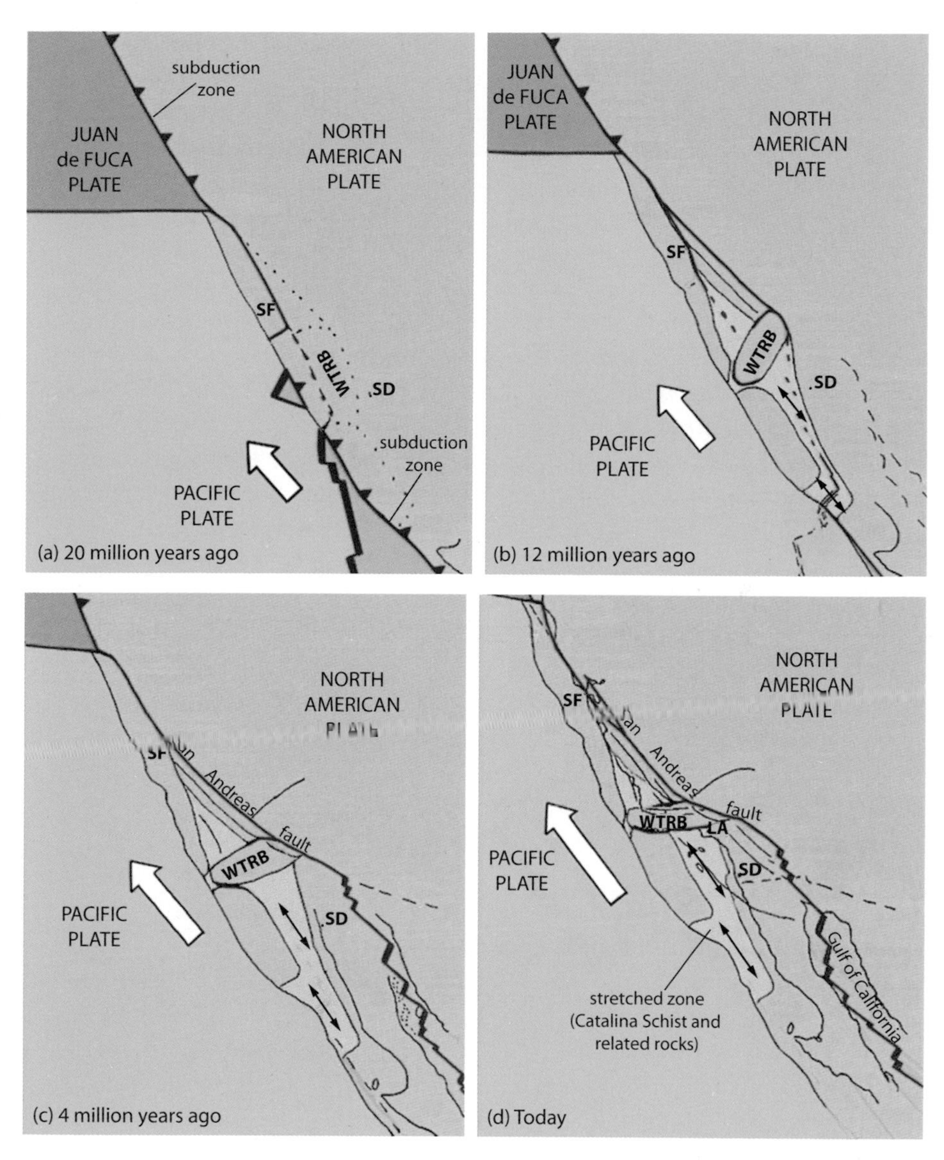

FIGURE 4.8. The tectonic evolution of Southern California during the past twenty million years has involved the serial capture of several pieces of the North American Plate by the Pacific Plate. WTRB = Western Transverse Ranges Block, SF = San Francisco, SD = San Diego, LA = Los Angeles. Red lines mark major faults and plate boundaries. Different plates are shown by different colors. (Modified from an animation by Tanya Atwater, University of California, Santa Barbara.)

rocks line up with equivalent rocks in San Diego, and all indicate flow west *toward* the ocean, which makes sense.

Perhaps the best evidence for the WTRB's clockwise spin as it emigrated from San Diego to where it is now comes from magnetic signals. When fresh lava chills, tiny magnetic minerals orient themselves toward magnetic north like miniature compasses. Once the lava solidifies, these mineral compasses become locked in place, recording the direction of magnetic north at the time the lava chilled. In lava beds that erupted *outside* the WTRB, the mineral compasses point generally north, telling us that the rocks lie more or less in the same orientation today as when they formed. But when we look at lava beds of the Conejo Volcanics that erupted *within* the WTRB, we see a strikingly different pattern. In lava beds fifteen million years old or older, the magnetic minerals point *east*. This makes sense only if the WTRB has spun clockwise by about ninety degrees since fifteen million years ago, so that the rocks have been turned from north to east, as shown in figure 4.9. Sediment layers of the Miocene-age Monterey Formation confirm this. Magnetic minerals in sediment layers, like those in lava beds, can record the direction of magnetic north at the time the layers formed. In Monterey Formation outcrops within the WTRB, the mineral compasses point generally east, just like in the lava beds. Moreover, when we look at lava beds and sediment layers younger than fifteen million years, we see that the mineral compasses point closer and closer to today's magnetic north as the rocks get younger, showing us the WTRB's gradual rotation to its present position.

To sum up, several lines of evidence—the match of WTRB rocks with equivalent rocks in San Diego, the direction that ancient rivers flowed, and the magnetic signals in lava beds and sediment layers—converge independently on the same conclusion. As the Pacific Plate dragged the WTRB northwest, the WTRB pivoted clockwise about ninety degrees, somewhat like a piece of airport luggage rotating when one end jams against the side of a baggage conveyor belt.

If you accept the evidence above, then you have to wonder: What happened in the gap left behind as the WTRB migrated northwest away from San Diego to where it is now? As the WTRB slid northwest, something had to take its place. Today, this region makes up the eastern half of the Continental Borderland, and most of it is underwater. (This is the area labeled "stretched zone" in figures 4.7 and 4.8.) But Catalina Island pokes up in the middle of this area, giving us a twenty-mile-long geologic sample that reveals what happened, we think, in the wake of

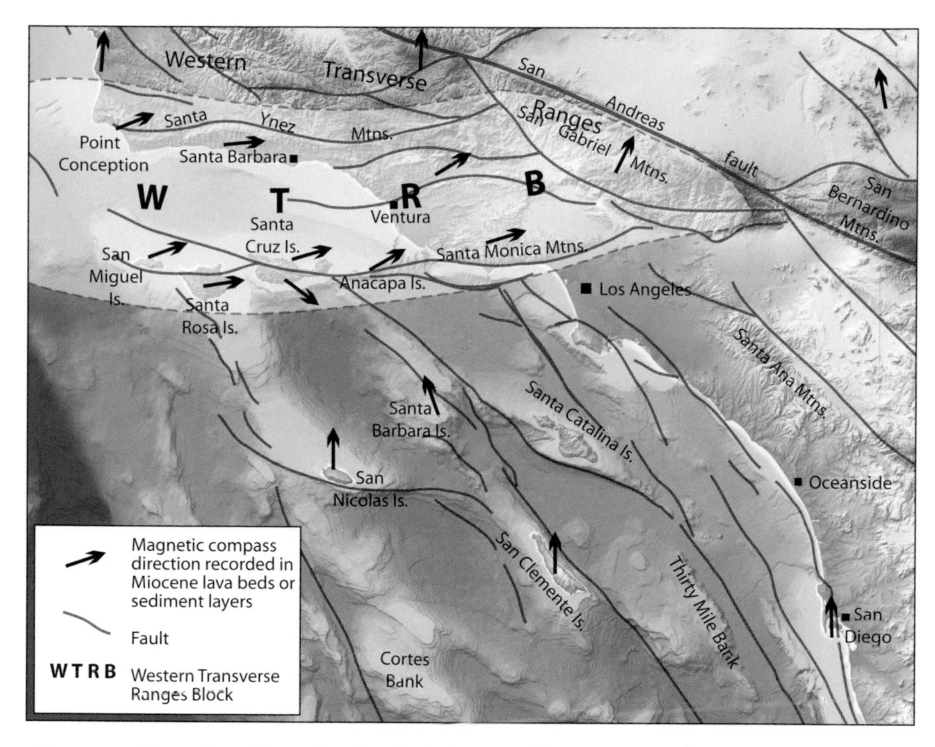

FIGURE 4.9. Magnetic evidence for the clockwise rotation of the WTRB (Western Transverse Ranges Block). The direction of ancient magnetic north is preserved in lava beds of the Conejo Volcanics and in sedimentary beds of the Monterey Formation, both of Miocene age. Each arrow marks a location where lava beds or sediment layers older than fifteen million years were sampled for remnant magnetism. Notice the east-pointing magnetic signal in rocks *within* the WTRB, compared to the north-pointing signal in rocks *outside* the WTRB. Clockwise rotation of the WTRB during the past fifteen million years explains the pattern, as shown in figure 4.8. (Shaded relief image from NOAA, with labels added; based on Luyendyk et al. 1985.)

the WTRB's northwestward migration across this area. Once again, here's the punch line up front: We think that, as the WTRB slid northwest, rock from deep in the old Farallon Plate subduction zone bobbed up to the surface to take its place. (By analogy, visualize, in ultraslow motion, a Jack-in-the-box popping up as you slide his lid off.) So what's the evidence that backs this up?

If the idea is right, it means that the rocks of Catalina Island (and, by extension, the rest of the inner Continental Borderland) once lay perhaps twenty miles or more *below* the WTRB, deep in the old subduction zone of the Farallon Plate. As the WTRB slid away to the northwest, these deep rocks popped up to the surface, probably because the weight of the WTRB above them was now gone, allowing them to rise buoyantly above the denser rocks around them. This may sound like a wacky notion, but it becomes less wacky when we explore curious features in the Basin and Range Province known as *metamorphic core complexes*. So here, we need to leave Southern California briefly and take a side-trip to the Basin and Range.

You might not think that the Continental Borderland has much in common with Nevada's Basin and Range, but their stories are linked in at least three ways. First, both regions formed over the same time span—roughly the past fifteen to twenty million years. Second, both formed (along with the San Andreas fault) as North America's western boundary shifted from subduction of the Farallon Plate to side-by-side sliding of the Pacific Plate past the North American Plate (a development portrayed in figure 4.6). Third—and most important—both the Continental Borderland and the Basin and Range have experienced profound *stretching*.

Geologists working in the Basin and Range have come to understand that large areas called *metamorphic core complexes* are key symptoms of this stretching. Picture a cat asleep under a stack of blankets. Slide the blankets off, and the cat rises and arches its back. The blankets represent young rocks—often layers of sediment or lava beds—that, as the Earth's crust stretches, slide away to expose old, deep rock beneath. This deep rock then arches upward buoyantly to occupy the space where the younger rocks above used to be. My own experience with metamorphic core complexes started as a geology graduate student at the University of Arizona in Tucson. Each morning, I woke to the sight of the Rincon Mountains east of town, and I often rode my bike up the Tucson Mountains west of town. The Tucson Mountains are made mostly of lava beds that once lay *above* the rocks of the Rincon Mountains. As the crust stretched east to west, the lava beds of the Tucson Mountains slid thirty miles to the west, and the deep rock beneath them bobbed upward to make the Rincon Mountains. In the cat-blanket analogy, the Tucson Mountains are the blankets and the Rincon Mountains are the cat. The whole system is a metamorphic core complex, and several dozen of them lie scattered throughout western North America, mostly in the Basin and Range. The term *metamorphic core complex* captures several meanings: *metamorphic* refers to rocks changed by heat and pressure deep underground; *core* refers to the nucleus of old, deep rock that rises buoyantly as the younger rock above slides away; and *complex* signifies the wide range of rock types and fault-tortured geology that occurs in these regions. Whether or not those details interest you, the take-home message is straight-up: *Metamorphic core complexes form where the Earth's crust undergoes widespread stretching.*

Returning now to Southern California, you may have guessed where I'm going with the Basin and Range analogy. We think that the rocks of that "stretched zone" labeled in figures 4.7 and 4.8 are much like the metamorphic core complexes of the Basin and Range. During the past few million years, as the WTRB slid away to the northwest, the rocks that once lay miles below—rocks deep in the old Farallon Plate subduction zone—bobbed up to the surface, just as they did in parts of the Basin and Range.

As mentioned, Catalina Island gives us our best geologic sample of the rocks in the stretched zone. But how can we tell whether or not the rocks on Catalina Island originally lay miles underground, deep in an old subduction zone? When rocks are buried deep in the Earth, they undergo dramatic changes related to heat, pressure, and fluid interactions—a set of processes known collectively as *metamorphism.* If we can figure out how rocks are metamorphosed when subduction drags them deep underground, we can compare that to what we find on Catalina Island. We don't have the technology to drill twenty or more miles down into, say, the Cascadia Subduction Zone (stops 1, 2, and 3 above) to see what's happened to the rocks there. But we can duplicate the conditions, on a miniature scale, using laboratory machines designed to torture rocks with intense heat and pressure.

We can start by torturing shale. Shale is simply solidified mud and clay, both of which collect abundantly on the deep ocean floor (as well as on lake bottoms and in mud puddles). A lot of shale surely rode on the Farallon Plate down into the old subduction zone. A subduction zone produces changes in rock known as *low temperature–high pressure metamorphism.* Pressures rise tremendously as the rocks are shoved down into the guts of the planet, but temperatures don't rise as fast with depth as they do elsewhere in the Earth because the cold subducting plate keeps the surroundings comparatively cool (although still scorching hot compared to temperatures at the Earth's surface). Shale subjected to low temperature–high pressure metamorphism yields a distinctive family of rocks called *schists,* illustrated in figure 4.10a. (Schists are a group of metamorphic rocks that show distinct layers or laminations; in geologic terminology, they are "foliated" (*folium* means leaf in Latin). Shove shale ten or more miles down a subduction zone, and the clay minerals within it transform into the blue-gray mineral glaucophane,

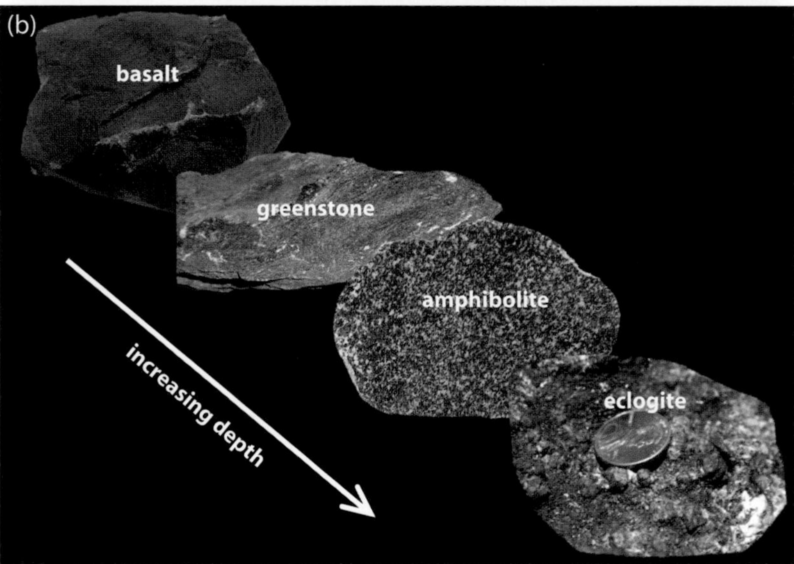

FIGURE 4.10. As seafloor rocks descend into a subduction zone, they experience *low temperature–high pressure metamorphism,* which means that pressures increase tremendously with depth, but temperatures rise much less rapidly than they do on average for the Earth. (a) If the starting rock is *shale* (a common sedimentary rock on the deep seafloor), the metamorphism goes from *blueschist* to *greenschist* to *amphibole schist.* (b) If the starting rock is *basalt* (the main volcanic rock of the deep seafloor), the metamorphism goes from *greenstone* to *amphibolite* to *eclogite.* The reddish mineral grains that you see in the eclogite are garnet minerals that sprout in the last stages of basalt metamorphism. (Rock images from Wikimedia Commons except the eclogite with penny for scale, photographed by Alex Barber.)

yielding an azure rock called *blueschist* that shimmers like the scales of a fresh-caught fish. Shove it down farther, and the glaucophane transforms into new minerals such as epidote, chlorite, and actinolite, all of which are green, thus yielding a rock called *greenschist*. Still farther down, these minerals convert into dark minerals like amphibole, forming *amphibole schist*. It's a marvel of geochemistry that a bland rock like shale, when cooked and squeezed in a subduction zone, gives rise to such a striking and beautiful range of metamorphic offspring.

How about if we torture basalt in a subduction zone? Volcanic pillow basalt paves the deep ocean floor, so surely a lot of basalt slid down the old subduction zone with the Farallon Plate. Low temperature–high pressure metamorphism of basalt yields its own suite of metamorphic offspring, different from those made by metamorphism of shale (figure 4.10b). In the first stages, chlorite minerals grow that tint the black basalt green, making *greenstone*. Farther down and under greater pressure, chlorite metamorphoses into black minerals like amphibole, yielding the rock *amphibolite*. Thirty or more miles down, where the pressures are one billion times greater than at the Earth's surface, garnet minerals sprout to form a rare and beautiful rock called *eclogite*. You could spend much of your life wandering the Earth and never see eclogite—and no wonder. It forms so deep underground that its trip to the surface is a rare event indeed.

Now for the payoff. Which of these offspring of low temperature–high pressure metamorphism in a subduction zone do we find on Catalina Island and elsewhere in the inner Continental Borderland? The answer is—*All of them!* All and more, in fact. Just about every rock that we can reasonably imagine plunging down a subduction zone has a representative on Catalina Island. Blueschist is the island's signature rock. It happens to be my favorite rock in California, not just because of the extraordinary tale it tells of burial and rebirth, and not just because it's irresistibly gorgeous, but because I get to use my special beer koozie when I photograph it (figure 4.11). But as much as I love blueschist, eclogite takes the prize for sheer profundity. Lay your hands on Catalina Island eclogite, and you are witness to a mind-boggling amount of geologic work: the removal of perhaps *thirty vertical miles of rock* that once lay overhead. What a story the rocks of Catalina Island tell. They were born in the dungeons of the planet, and they would have stayed there, forever hidden, if not for the door that opened above them—the door of the northwest-migrating WTRB—which released them from their Hadean home, and let them pop upward to see the California sun.

FIGURE 4.11. An outcrop of the Catalina Schist near Two Harbors on Catalina Island, with my special beer koozie ("Schist happens so have a Gneiss day") for scale. The rock is blueschist, a signature rock produced by low temperature–high pressure metamorphism of shale in a subduction zone (see figure 4.10a). The beer is Budweiser, a light-flavored and nonfilling beer suitable for intense days of geologic field work. (Photograph by the author.)

One coda will complete this story of Southern California's geologic birth. Today, Catalina Island sheds boulders from its flanking cliffs into the thundering surf. Scuba diving around the island, I've seen many of the deep-subduction-zone rocks described above now forming boulders that shelter schools of Garibaldi and anchor forests of giant kelp. Anywhere you find piles of big boulders, it's reasonable to think that a source—a high, steep place from which the boulders tumbled—should be close by. That's where the conundrum of the San Onofre Breccia comes in.

Breccia (sounds like "fetch ya") is a geologic term for a sedimentary rock made up of big, sharp-edged boulders. Take a sample of bedrock from Catalina Island, bust it into sharp-edged pieces ranging in size from a fist to a washing machine, throw them in a huge pile, and you

FIGURE 4.12. An outcrop of the San Onofre Breccia at Aliso Beach in Orange County. We think that most rocks within the San Onofre Breccia were formed by low temperature–high pressure metamorphism deep in the subduction zone of the Farallon Plate. Later, they were uplifted and eroded from now vanished highlands to form the bouldery deposits of the San Onofre Breccia. The eraser end of the pencil lies on a boulder of blueschist, and the pencil point touches a chunk of greenschist. Both are common products of low temperature–high pressure metamorphism, as shown in figure 4.10. (Photograph by the author.)

have a likeness of the San Onofre Breccia (figure 4.12). The San Onofre Breccia is widespread in Southern California; we find patches of it along the mainland coast from near Tijuana all the way to the Santa Monica Mountains, as well as offshore on Anacapa, Santa Cruz, and Santa Rosa islands. The bouldery beds can stack up hundreds of feet thick and contain rocks as big as grizzly bears. Fossil shellfish within the San Onofre Breccia pin its age at fourteen to seventeen million years. Apparently, by about seventeen million years ago, the metamorphosed rock that once lay deep beneath the WTRB had bobbed upward high enough to form highlands that loomed across large portions of the Continental Borderland—highlands that looked, perhaps, like ancient versions of

Catalina Island. These highlands shed boulders off their flanks to form the San Onofre Breccia.

I say "ancient versions of Catalina Island" because Catalina Island itself could not have been the source of the San Onofre Breccia. For one thing, the island is too young. (It was pushed up out of the sea within the past few million years by the big restraining bend in the Catalina fault that you can see in figure 2.4b.) For another, the San Onofre Breccia is too widespread to have been shed from a single highland source. So we're forced to one conclusion: A whole bunch of Catalina Island–like high places must have bobbed skyward early in the formation of the inner Continental Borderland. These highland sources shed boulders thickly across Southern California between about fourteen and seventeen million years ago. Then, as the crust continued to stretch in the wake of the WTRB's ongoing northwestward migration, these highland sources apparently foundered below the sea. Drilling and dredging today beneath the seabed of the inner Continental Borderland, we find Catalina Island–type rocks everywhere. It's fun to imagine that some of these must once have soared high above the ocean, turning breakers white against their flanks as boulders bounced down from their mighty heights to splash into the Miocene sea.

If you've made it this far, the beer is on me. Learning the bedrock story of Southern California in one sitting—even a stripped-down version like this one—is no easy task. Many professionals still struggle to put all the pieces together. If this chapter had been about the geology of, say, Iowa, it would have been easier. In the baseball movie *A League of Their Own,* the coach (played by Tom Hanks) berates one of his players for complaining that the game is too hard. "It's supposed to be hard," he yells. "The hard is what makes it great." California geology is hard. The state is a geologic train wreck of rocks, faults, and moving plates, still growing in tectonic adolescence. This chapter sums up prevailing interpretations of present evidence, but we don't know whether we've got the whole story right. If the history of science teaches us anything, it's to never be too sure of ourselves. New discoveries regularly change prevailing wisdom and occasionally overturn it completely. Already, a new generation of scientists—swinging rock hammers from islands and mountaintops, and digging up samples from the seabed—is unearthing tales no one has yet imagined.

That's what makes it great.

5

Waves and Surfing

As it reaches its uppermost limit [on the beach] the wave dies; all the energy so carefully gleaned from the winds of the distant storm and hoarded for a thousand miles of ocean crossing is gone, expended in a few wild moments.

—Willard Bascom, *Waves and Beaches*

There is a place a hundred miles west of San Diego where, if the ocean is calm and the tide low enough, you can stand in waist-deep water on the summit of an immense seamount. There's no land visible. The nearest island, San Clemente, is forty-five miles southeast and likely shrouded in fog. A few miles in every direction, the ocean floor plunges more than five thousand feet. Deep currents run into the seamount, carrying nutrients up to the sunlit surface waters. The nutrients—dissolved nitrate, phosphate, and iron mostly—act like garden fertilizer, fueling the photosynthetic growth of tiny phytoplankton and forests of giant kelp. These form the base of a rich food pyramid that, via endless slaughter, ascends from zooplankton to fishes small then large, to seals and sea lions, and finally to great white sharks, who patrol the seamount with the insouciance of apex predators. The ocean is full of teeth here. But the bravest and most adrenaline-addicted members of the big-wave surfing community can't resist this place. This is Cortes Bank, home to some of the largest surfable waves on Earth.

Cortes Bank is not a welcoming place: gusty and cold, frequently swept by treacherous currents and titanic swells (large storm-generated waves), and often thick with fog. If not for the clang of a Coast Guard marker buoy near Bishop Rock (the barely awash summit of the seamount), you might think yourself adrift in a world before humans—a Jurassic ocean, perhaps, haunted by long-necked plesiosaurs. Until recently, Cortes Bank remained virtually unknown except by a handful of abalone fishermen and scuba divers, some of whom told of seeing colossal breaking waves there. On January 23, 1990, two surf-seekers named Larry Moore and Mike Castillo decided to see what the big-wave rumors were about. A violent Aleutian storm had recently kicked up monster swells that were bombarding Hawaii and California with legendary surf. Moore and Castillo took off from Oceanside Airport in a small plane and aimed it west–southwest at Cortes Bank. Passing San Clemente Island, which had been dampening the swells, the two men stared slack-jawed at the outsized corduroy pattern of the ocean. "The lines of swell were unbelievable. The interval on the swell was just huge," Castillo remembered. A Navy ship they flew over "just looked like a toy boat in those waves." Ten minutes later, they spied whitewater. Gigantic waves were unloading onto Bishop Rock, burying the marker buoy in violent cascades. Castillo dropped the plane low, and they looked *up* as wall after cerulean wall rose from the corduroy ranks to crest and tip forward at . . . sixty feet high? . . . eighty feet? It was hard to tell, but "if you surfed down there, there was a serious chance of death or dismemberment," Castillo decided. "It looked like nothing anywhere else, even Jaws over on Maui."*

Cortes Bank runs about twenty-five miles long and seven miles wide. Its long axis points northwest. Lay a ruler on that long axis, follow it two thousand miles northwest, and you'll find yourself in the center of the storm-wracked North Pacific—one of the world's great wave factories. The alignment of Cortes Bank happens to be perfect for taking big swells cranked up by North Pacific storms and focusing them, like light rays through a magnifying glass, so that they bloat into azure monsters as they cross the bank.

Surfers rightly say that no two waves are the same. But the physics behind waves and surf is much the same everywhere. After we work through some principles of wave science, we'll go visit six classic Southern California surf sites, including Cortes Bank (figure 5.1). I hope to show you that, whether it's sixty feet at Cortes Bank or six feet at Rincon Point, classic surfing waves form for the same basic reasons everywhere.

WAVE BIRTH

The life of a typical surfing wave begins in a storm and ends a few days later when the wave breaks on a beach, shoal, or seamount. During its

* Quoted material throughout this chapter is cited as told to Chris Dixon in his book *Ghost Wave* or is derived from interviews on www.surfline.com. For details, see Notes on Sources.

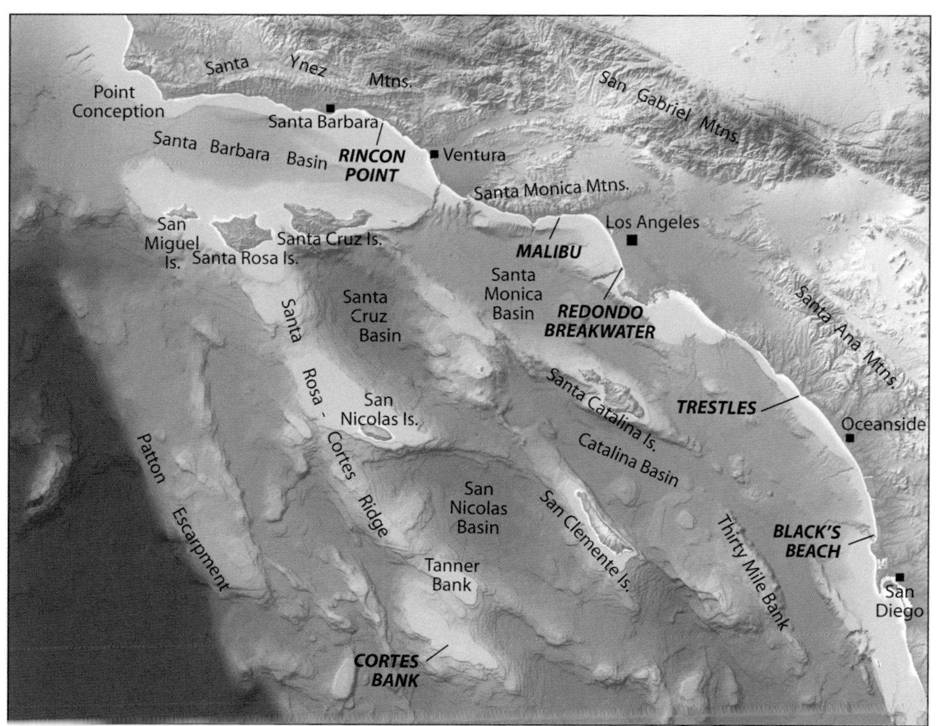

FIGURE 5.1. Six classic Southern California surf sites that we visit in this chapter. Under the right conditions, ocean swells and bottom topography conspire to produce spectacular surfing waves at each place. (Shaded relief image from NOAA, with labels added.)

short lifetime, a wave obeys unvarying laws of physics, allowing us to predict its behavior with fair accuracy. I sometimes hear surfers speak of waves as if they were fickle, mysterious things. I think the feeling comes from their beauty, symmetry, and ceaseless motion—features we associate more, perhaps, with wild animals than with the physical world. But wave behavior is as predictable as gravity or planetary orbits. You just have to know the variables.

Most ocean waves are formed by wind, and three factors control the size of waves that the wind can make: wind speed, duration, and fetch (the distance of water over which the wind blows). The faster the wind, the longer it blows, and the greater the fetch across which it blows, the larger and faster the waves. Local shore breezes don't make worthwhile surfing waves because they don't have enough speed, duration, or fetch. The breeze blowing your hair at the beach didn't make those waves you see the surfers riding. Local breezes make *chop:* tiny waves that wrinkle

Far-Traveled Swells

Even though swells from Pacific storms dominate the California coast, some of the waves you see breaking on our shores begin as far away as the Indian Ocean. We know this thanks to the work of Walter Munk at the Scripps Institution of Oceanography. Different waves produced by a single storm will travel at different speeds—a process called *dispersion*. The farther away the storm, the greater the time gap between the arrivals of faster versus slower waves from that same storm. Munk's calculations from these time-gaps showed that some swells arriving in Southern California must have started in storms more than ten thousand miles away, which meant these waves formed outside the Pacific Ocean! The source turned out to be the southern Indian Ocean, halfway around the world. These swells make their long journey from the Indian Ocean to California by passing through the gap between Antarctica and Australia. You can convince yourself of this by stretching a string across a globe (or a virtual string on Google Earth). You will find that you can draw a straight line from the southernmost Indian Ocean to California without bumping into land—one of the longest journeys for waves anywhere on Earth.

the surfaces of the big swells. On a gusty day, the chop might reach one or two feet—big enough to make a mess, but no good for surfing. To make good surfing waves, you need storms that blow for several days across wide stretches of open ocean to make *swells:* large, storm-generated waves that may travel for thousands of miles before breaking on your local beach.

The reason the Pacific Ocean hosts so many world-class surf sites comes down to fetch and storms. The fetch across the Pacific is second only to that in the Southern Ocean (the continuous belt of water wrapping around Antarctica, home to the biggest open-ocean waves on Earth). As figure 5.2 shows, three major Pacific storm centers send swells to Southern California: the North Pacific (where storms form mostly from October through May), the tropical Pacific (which spawns hurricanes from June through October), and the South Pacific near Antarctica (most active from April through November—the Southern Hemisphere winter). Although swells from the North Pacific dominate California's coast, swells coming to the Southern California Bight from that direction (the northwest) are dampened by Point Conception, which sticks into the ocean like an elbow to block northwest swells.

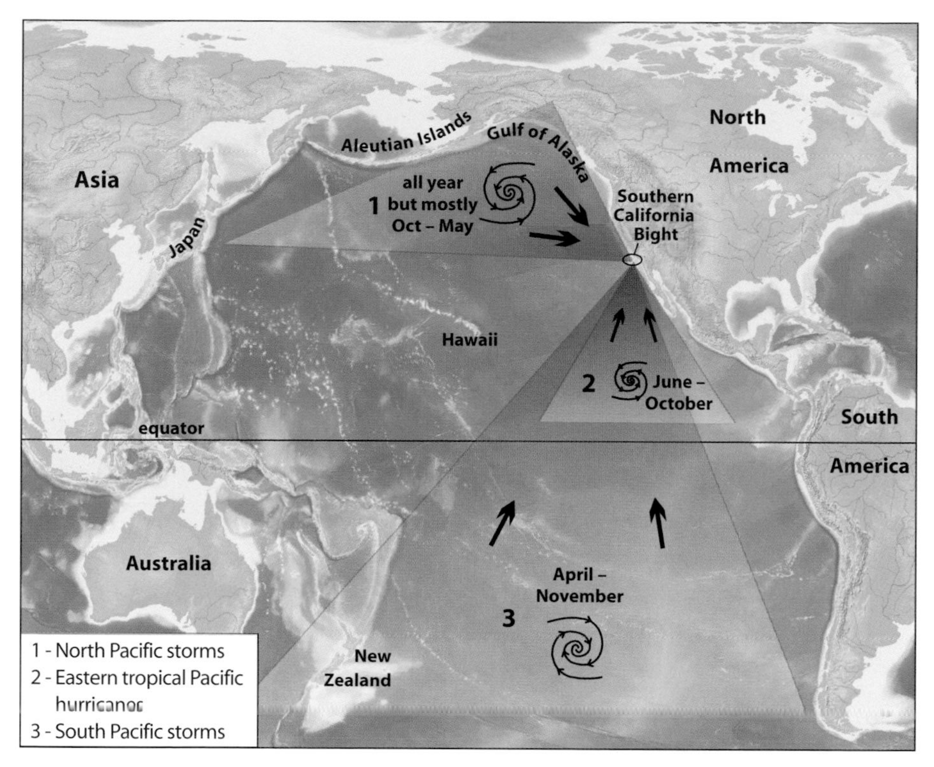

FIGURE 5.2. Major storm centers and swell sources for the Southern California Bight. The spiral symbols represent cyclonic storms—the source of large ocean swells.

(Figure 1.11 shows an example of Point Conception's blocking effect.) By contrast, swells approaching the bight from the tropics or the South Pacific are dampened by only a few offshore islands. South swells are thus an important source of surf breaks in Southern California, especially in summertime.

Whatever their source, it's always fun to think about the journey of ocean swells, some of which may have traveled nearly halfway across the planet to give you a ride during the last moments of their existence (see text box). In some ways, the journey is far greater than that. A big swell arriving in Southern California represents the energy of solar nuclear fusion that has traveled ninety-three million miles from the Sun to heat up the atmosphere and ocean, spawning a pinwheeling low-pressure system whose fierce winds have converted the Sun's heat into big waves. When you ride far-flung ocean swells, you're grabbing, for a few seconds, your own little piece of solar fire.

WAVE PERIOD: THE ESSENTIAL NUMBER

No measurement is more important for understanding waves and surf than the *wave period:* the time, in seconds, between wave crests as they pass a stationary point. Visualize standing on a pier with a stopwatch as swells roll by beneath you. When one passes, you start the watch, and when the next passes, you stop it. Collect a bunch of readings, average them to even out measurement error, and you have the period. My students do this every semester on a local pier, and the periods they get using two-dollar stopwatches are usually spot-on with those obtained by high-tech automated wave buoys.

The wave period is important for three reasons:

1. It directly reflects the amount of energy that waves carry; longer-period waves carry more energy.
2. It lets us predict how waves will bend and concentrate their energy at particular places along a coast to make big surf.
3. Dozens of automated offshore buoys measure wave periods continually, and the data are widely available on the Internet.

Figure 5.3 shows that waves increase their periods in direct proportion to the amount of energy that storm winds put into the ocean. The greater the wind speed, duration, and fetch, the longer the period. Waves with longer periods travel faster, have longer wavelengths (meaning the distance between each wave crest), and cause the water to move at greater depths. Notice in figure 5.3 that the water doesn't travel across the ocean surface with the wave. Instead, the water moves in a circle called a *wave orbit,* ending up nearly back where it started as each wave passes. The longer the wave period, the larger the wave orbits, and the deeper they reach below the surface. The *wave base* is the depth where this orbital motion stops, and it increases in direct proportion to the wave period. As we'll see, the wave base is one key to understanding how and where good surfing waves form.

If you scuba dive, you may have felt wave orbits below the surface. When you first submerge, you feel your body tracing a lazy circle as the waves pass by overhead. The deeper you go, the smaller the circles, and once you drop below wave base you don't feel any motion at all. Some years ago, I was diving on a research project in southern Baja California. Big hurricane-generated swells were cruising by overhead, and as my dive partner and I dropped to the seabed at forty feet, we found it impossible to stay put over our observation stations. The wave orbits, flattened against

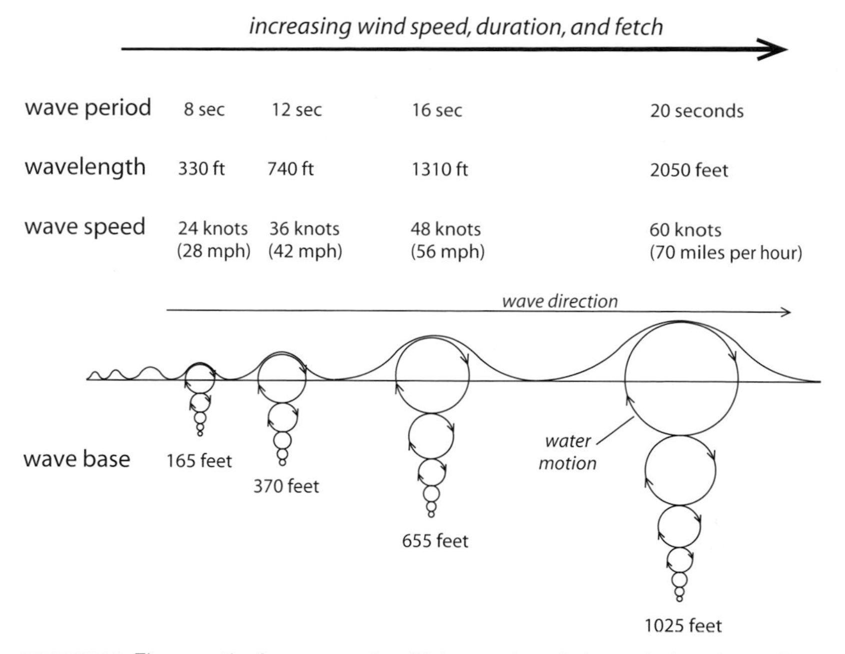

FIGURE 5.3. The growth of ocean swells with increasing wind speed, duration, and fetch. Wave period, wavelength, wave speed, and wave base all increase as the wind transfers more of its energy to the ocean.

the seafloor, swept us first one way and then the other, returning us to the same place every seventeen seconds—a cycle that precisely matched the wave period. We resurfaced, collected every pound of extra lead weight we had brought, and dumped all the air from our buoyancy compensators, effectively turning ourselves into stones. Wedged among big boulders, we managed to get some work done in the back-and-forth surge—until suddenly the surge dislodged my partner and swept her away. The surge brought her back a few seconds later, flying toward me with limbs akimbo. We grabbed hands, and she managed to wedge herself back into the shelter of the boulders. Braced more firmly now, we got our work done amid reversing hailstorms of sand grains and swim-bys of curious fish.

Because waves with longer periods travel faster, waves spawned by a storm will spread away unevenly, with the faster, longer-period waves moving out ahead of the slower, shorter-period ones. It's a bit like cars leaving a stoplight; the faster cars get ahead of the slower ones, and so the cars become more spread out farther down the road. This process is called *wave dispersion,* and it transforms the way the ocean looks. In a storm, the sea is a chaotic mess of many waves with varying periods, speeds, and wavelengths. But let the waves travel away from the storm for a day or two, giving them time to disperse, and you'll see an ocean surface that looks like a washboard, made up of equally spaced waves with the same period, speed, and wavelength that we call *swells*. When these storm-generated swells approach land, "Surf's up!"

Or not. To form good surf, something else has to happen to storm-generated swells as they approach the shore.

GETTING TO THE BOTTOM OF A GREAT WAVE

Almost everyone appreciates a good bottom, but in surfing, the idea takes on special meaning. "It's hard to get an epic wave without an epic bottom," the surf forecaster Sean Collins has remarked. Earlier, I introduced the concept of *wave base:* the depth of water where the orbital motion of a passing wave dies out. Swells moving through water deeper than their wave base don't feel the bottom. But once swells come into water shallower than wave base, the orbital motion of the waves begins to feel the seabed, and the waves slow down.

You might assume that waves slow down in shallow water because of friction against the seabed. It's a widespread misconception, probably because it's intuitive to think of waves dragging across the seabed and being sapped of energy as they travel toward shore. But measurements show that this is not the case. Waves approach shore with almost no loss of energy, hoarding it right up to the moment that they break. We can sidestep many details with one simple realization: The wave period *does not change* as waves approach the shore, and, as I explained above, the wave period is directly proportional to the amount of energy that a wave inherits from the wind. So, if waves don't lose energy as they approach shore, why do they slow down? The answer is related to conservation of energy, a basic law of physics. When the orbits of approaching waves make contact with the seabed, their motion changes in such a way as to push up the wave height. The amount of energy carried by a wave can be expressed mathematically as wavelength multiplied by wave height squared. This means that if wave height increases even a little, wavelength must decrease a lot to keep the total energy the same. As the wavelength decreases, the wave slows down, because wave speed equals wavelength divided by wave period, and the period does not change. Bottom line: The total energy carried by a wave remains constant even as it slows down in shallow water because *the energy is converted into increasing height*. This slowing and growth in height leads inevitably to every wave's demise. When the height:length ratio of a wave reaches about 1:7 (as it must, since the height always rises and the wavelength always shortens as a wave enters shallow water), the wave topples forward and breaks. In those few seconds, all the energy gathered from a distant storm and conserved across many miles of ocean converts into a forward-pitching cascade. If you're a surfer who has picked your spot right, that's the energy that takes you for a ride.

Whether or not these details interest you, the key point is this: *The deeper the wave base, the sooner that swells approaching shore will begin to feel the bottom and slow down.* As figure 5.3 shows, the wave base for any wave is equal to exactly half the wavelength. So, for example, a swell with a five-hundred-foot wavelength will begin to feel the bottom and slow in two hundred fifty feet of water, whereas one with a two-hundred-foot wavelength will begin to slow in one hundred feet of water. However, it's not easy to accurately measure wavelengths on moving ocean swells. So instead, if we want to know the wave base, we use that essential number: the wave period. Dozens of automated buoys measure wave periods continuously throughout the Pacific and other ocean basins, allowing us to calculate the wave base with a simple equation:

wave base (in feet) = 2.56 × period (in seconds) squared

From the equation, a swell with a ten-second period will have a wave base of (2.56 × 10 × 10) = 256 feet, whereas a swell with a twenty-second period will have a wave base of (2.56 × 20 × 20) = 1,024 feet. Notice that, because wave period is squared in the equation, *doubling* the wave period *quadruples* the wave base. This means that even a small increase in wave period translates to a large increase in wave base. This gives us another key point: *Long-period waves have deeper wave bases and will thus begin to slow in deeper water than short-period waves.*

What, you may wonder, is so important about waves slowing down as they enter shallow water? The answer is that the more waves *slow*, the more they can potentially *bend* around bottom topography and concentrate their energy to form good surfing waves.

A couple of terms will help me explain this idea. *Bathymetry* means the shape and depth of the ocean bottom. *Refraction* refers to how waves bend and change direction as they meet various types of bathymetry. Figures 5.4 and 5.5 show how swells will refract over two common types of bathymetry: an undersea ridge and an undersea canyon. Waves always bend toward shallower water, conforming somewhat to

Long-period swell over an undersea ridge
wave period = **16 seconds**, wave base = **650 feet**

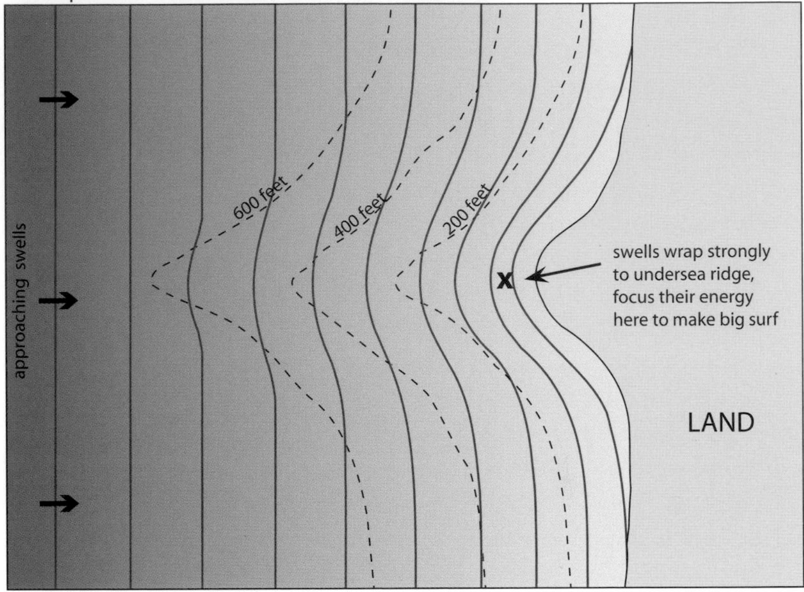

Short-period swell over an undersea ridge
wave period = **8 seconds**, wave base = **165 feet**

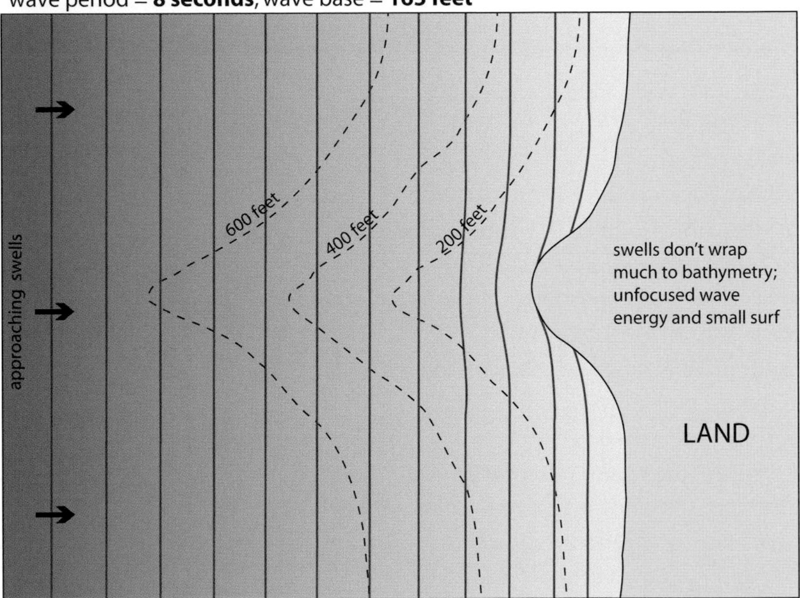

FIGURE 5.4. The effect of wave period on refraction of ocean swells over an undersea ridge. Long-period swells (upper image) feel the bottom in deep water and wrap toward the ridge, focusing their energy to form big surf at the spot marked X. Short-period swells (lower image) don't feel the bottom until close to shore. Their energy remains unfocused, and they form only small surf.

Long-period swell over an undersea canyon
wave period = **16 seconds**, wave base = **650 feet**

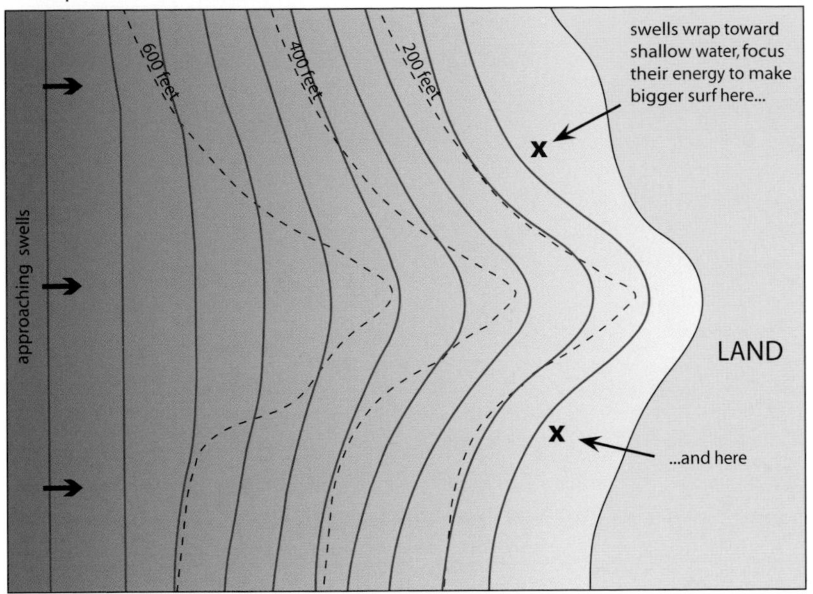

Short-period swell over an undersea canyon
wave period = **8 seconds**, wave base = **165 feet**

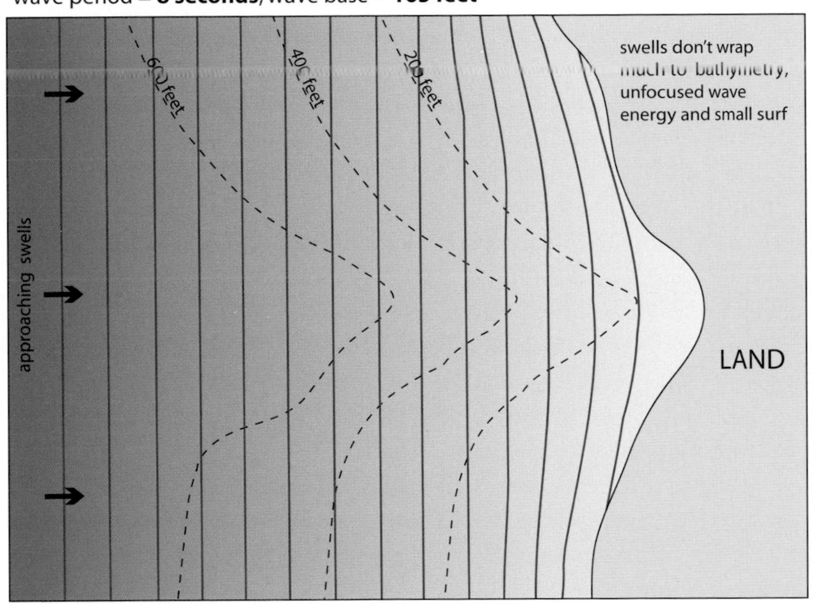

FIGURE 5.5. The effect of wave period on refraction of ocean swells over an undersea canyon. Long-period swells (upper image) feel the bottom in deep water and bend away from the canyon, focusing their energy at the two spots marked X. Short-period swells (lower image) don't feel the bottom until close to shore. Their energy remains unfocused, and they form only small surf.

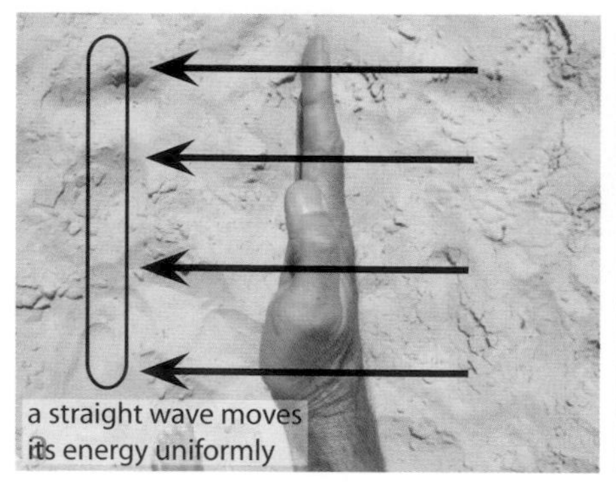

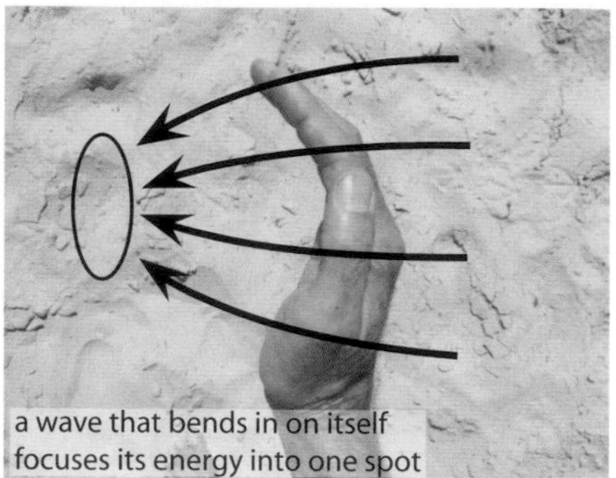

FIGURE 5.6. The best surfing waves often form where waves refract (bend) concave as they approach shore. Such bending focuses the waves' energy into a small area, making them bigger and more powerful when they break.

depth contours. But the thing that's most important to notice in figures 5.4 and 5.5 is how the long-period swells (i.e., those with deeper wave bases) start bending sooner than the short-period swells, and therefore bend more, conforming or "wrapping" more closely to the depth contours.

In general, the more that swells bend in response to the shape of the bottom, the more they can potentially focus their energy into certain spots to make good surfing waves. The key is to bend swells in such a way as to focus their energy into a small area. Hold your right arm out with your fingers straight, as if you were going to shake someone's hand, like the left image in figure 5.6. Sweeping your hand in the direction of the arrows mimics the straight line of a moving swell. Now curve your palm and fingers as in the right image. That shape—concave in the direction the swell is traveling—is an ideal way to make good surfing waves. When the bottom bends swells into that shape, it focuses their energy into the center—represented by the middle of your curved hand—forming bigger, more powerful waves. Undersea ridges and undersea canyons are particularly good at focusing wave energy this way, although in different ways (figures 5.4 and 5.5).

I've thrown a fair amount of surf science at you in these last few pages, so let me sum things up before we put these concepts to work as we explore surfing in Southern California. Call these my *Four Essential Concepts of Surf Science:*

1. Storm winds make most large ocean swells. The longer the speed, duration, and fetch of the wind, the more energy is transferred to the ocean to make waves.

2. More wind energy produces swells with longer periods, longer wavelengths, higher speeds, and deeper wave bases.
3. The wave period is the most useful number for forecasting how swells will behave as they approach shore. The longer the wave period, the sooner that approaching swells will begin to feel the bottom, causing them to refract (bend and change direction) more in response to the shape of the bottom.
4. Where refraction causes swells to bend concave in the direction of travel, their energy will focus to produce bigger waves. Undersea ridges and undersea canyons are both good at focusing swells this way and are thus responsible for the waves at many classic surf sites.

CORTES BANK REVISITED: RIDERS BETWEEN THE STORMS

New Year's Day, January 2008. A huge low-pressure system off the Aleutian Islands is vacuuming in air from across a half-million square miles. The storm soon bloats into a gargantuan pinwheel that takes up most of the North Pacific. The pressure gradient between low and high is so extreme that the barometric pressure lines bunch together like cliff-contours on a topographic map. Small boats flee for the nearest harbors, and captains of oceangoing freighters plot wide routes around the rising fury. On January 3, the storm slams into Washington and Oregon. It reaches San Francisco the next day, ripping up hundred-year-old oaks as if they were garden weeds and dismantling the power grid to plunge 2.1 million people into darkness. Ten feet of snow fall in the Sierra Nevada, where weather stations record wind gusts of 165 miles per hour. The storm rakes south to San Diego, chased by a second spiraling storm, slightly less powerful. The gap between them brings up a question: Could someone find surfable waves in the brief calm between the two storms?

Rolling outward from the storm, the swells look like outsized versions of the expanding rings made where you toss a rock into a pond—only this pond is the Pacific Ocean, and the rings will reach New Zealand in a week. Their heights decrease as they spread out, but even after several thousand miles they are an awesome sight: twenty-three feet high from trough to crest, rolling along at about seventy miles per hour, with twenty-second periods and wavelengths of more than a third of a mile. They are some of the most powerful Pacific waves of the decade:

moving mountains of pure, long-period energy, destined to explode somewhere.

Traveling through deep water, the swells move unimpeded by the ocean floor and thus hold their shape and speed. Their orbital energy reaches down a thousand feet, but the open Pacific averages more than two miles deep, so the waves don't feel the bottom. Approaching the shallowing waters along the mainland California coast, the swells rise up into colossal surf that the storm winds immediately shred to pieces. But at Cortes Bank, things are different. The swells arrive, on January 5, in the narrow window of calm air between the first and second storms.

A single boat arrives the same day, just after noon. It is a thirty-six-foot Twin Vee 550-horsepower catamaran owned and piloted by surf photographer Robert Brown. His cargo, loaded before dawn at Dana Point Harbor, includes two jet skis and four surfers: Mike Parsons, Brad Gerlach, Greg Long, and Grant Baker. All are at the pinnacle of the big-wave-surfing profession, regularly winning top awards. In January 2001, Mike Parsons set a record at Cortes Bank by riding a sixty-six-foot wave. But today is different. "Compared to the wave Mike had out there in 2001, this was so much bigger," Greg Long remembered. "I saw a lot of eighty-foot-plus waves. It was the most incredible thing I've ever seen in my life."

Throughout the Continental Borderland, faults have raised some blocks of the Earth's crust to make islands and shallow banks, and dropped other blocks to form deep basins. A few extra feet of uplift, and Cortes Bank would be an island today. In fact, it *was* an island, more than three hundred feet above the sea, some twenty thousand years ago, at the peak of the latest ice age. As the Earth warmed and ice up to two miles thick melted away from Canada and northern Eurasia, the rising sea submerged Cortes Island. (We'll come back to the story of the ice ages in chapter 7.) Today, big swells approaching from the northwest encounter a sea bottom that rises, in just a few miles, from five thousand feet deep up to a broad, shallow ridge perfectly aligned to impale the swells and bend them inward over the bank. Figure 5.7 shows my expectation of how a twenty-second period swell from the northwest will refract over Cortes Bank. You can see that the greatest wrapping—and thus the largest surf—will occur as the waves approach the shallowest part of the bank near Bishop Rock.

Bobbing over huge, butter-smooth swells in deep water south of Bishop Rock, the men on Rob Brown's boat stare thunderstruck at the

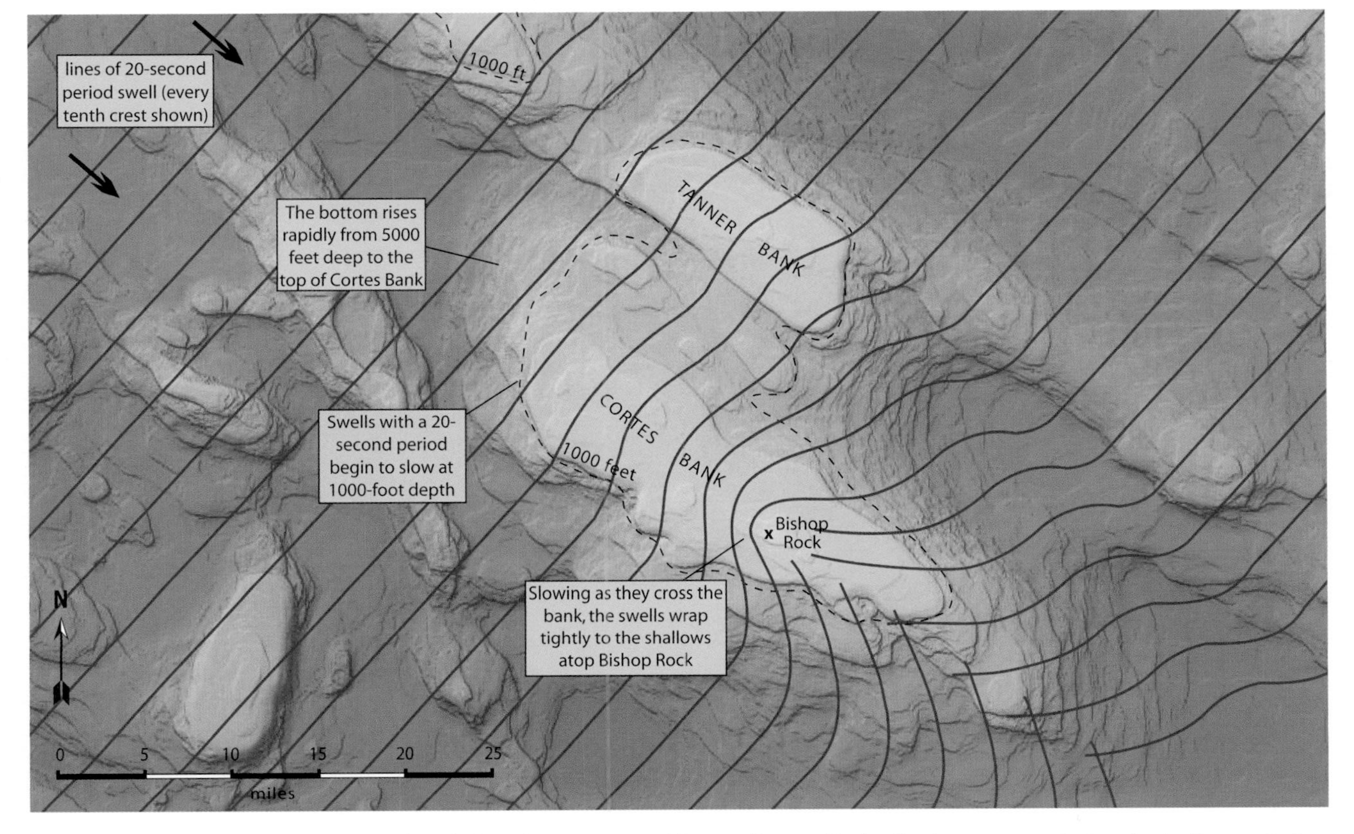

FIGURE 5.7. Refraction of twenty-second-period northwest-approaching swells over Cortes Bank, a seamount a hundred miles west of San Diego. At the scale of the diagram, the blue lines represent every tenth wave crest. The wrapping and focusing of big northwest swells over the bank can create breaking waves more than a hundred feet tall near Bishop Rock, the peak of the seamount. (Shaded relief image from NOAA, with labels added.)

scene to the north. Waves eighty feet high or higher rear up and topple over, squirting geyser-like explosions of whitewater nearly twice as high into the air. "You could count the seconds from when you saw one throw out as it fell," cameraman Matt Wybenga remembered of the waves. "One thousand one, one thousand two, one thousand three, one thousand four—till it detonated. It was so loud." As they curl over, the waves create almond-shaped tubes big enough to hold a six-story building. They were "these giant, giant tubes," Brad Gerlach recalled, and "you'd look at it and you're like, *maybe I could ride that.* Then it would clamshell and explode, and you're like, *oh no, OH NO.*" The waves have stirred up a terrifying cauldron of whitewater around Bishop Rock, and have heaped so much water onto the mesa-like summit of the rock that the ocean pours off in all directions, as if off a table. As the big waves collapse, they sling forth smaller waves that charge through the whitewater mess. It all adds up to a perfectly murderous scene—the last place on Earth you would want to be, but where a surfer might end up if he wipes out. "If you lost your guy in there, he was just *gone,*" Grant Baker remembered thinking. "He would have been lost in that expanse and you would never find him. It was just so scary."

As the swells rise up onto Cortes Bank, they slow from their deep-water speed of roughly seventy miles per hour to less than forty as they crest and begin to break. Such speeds—far faster than anything most of the world's surfers will ever experience—demand tow-surfing, where the surfer, his feet inserted tightly into straps on his board, is slung into the wave by his partner on a jet ski. At first, the teams are cautious, with the surfers catching the far right shoulders of the waves and staying well away from sixty-foot-high tubes detonating to the north. The trick, of course, is not to fall. As big waves break, they take so much air down with them that the water's density drops to something near that of beer foam. A body that will float in water will sink in foam. This, along with the waterfall-like force of the breaking wave, produces what surfers call a *hold-down,* which can trap you in swirling blackness for a minute or more. On top of this, of course, is the risk of being swept into the murderous whitewater near Bishop Rock.

As the afternoon wears on, the winds settle in the peak of calm between the storms, and the hazardous chop crossing the faces of the big waves flattens out. The surfers begin to ride with more confidence. Mike Parsons catches a sixty-five-footer and rides it more than a mile. Soon after, a behemoth begins to stand up to the west. Brad Gerlach, driving the ski, slings Parsons in at full throttle. Parsons plummets down the face, going faster than he ever has on a surfboard—so fast, in fact, that a frightening turn of physics takes over. The fins on his board are slicing the water so quickly that pockets of pure vacuum, known as cavitation bubbles, form around them, creating drag. It's like pulling an emergency brake, and now Parsons is moving slower than the wave, which is rearing and lifting him from behind. In relation to the wave, he's going *backward* up the face. Closer and closer to the crest he goes, his angle of descent steepening to nearly straight down. He's nearly at the lip—the worst place to fall. Whitewater begins to chandelier down on him. "It's gonna hit you, but you gotta make this. Point it. Just stay on. You can't fall. You can't die," Parsons thinks. Suddenly, the cavitation bubbles release. Parsons rockets forward just as the tube begins to collapse, and aims for safety along the wave's shoulder a half-mile away.

Parsons's wave, officially measured at seventy-seven feet, became a world record at the time. But Parsons concedes that his was not the biggest wave ridden at Cortes that day. That distinction went to Greg Long, who rode (unphotographed, and thus unofficial) a wave that was easily eighty feet and perhaps ninety feet high. "I guarantee you there will be a hundred-foot wave ridden out there," Parsons says of Cortes Bank. "If you put yourself in the right place at the right time, it will happen."

SURF ON THE MAINLAND COAST

Record-making waves and record-breaking elite surfers are fine, but what about mainland Southern California, where mostly normal people ride mostly normal-sized waves? Even during the biggest swells, surf along the mainland never comes close to the size of surf at Cortes Bank, for two reasons. First, the Southern California Bight faces generally southwest and thus receives big northwest swells indirectly. Second, the islands and banks of the Continental Borderland block and dampen approaching swells. These two constraints don't apply north of Point Conception, which is why waves are consistently higher there. (At Mavericks, near Half Moon Bay south of San Francisco, an undersea ridge amplifies long-period northwest swells in much the same way as the bathymetry at Cortes Bank, and waves there can approach the size of those at Cortes Bank.)

As we explore surf along the mainland coast, I'll tap back into the *Four Essential Concepts of Surf Science* that I listed a few pages back, and add one more:

5. The direction of approaching swells exerts strong control on surf along the Southern California mainland because of swell shadows and swell windows formed by offshore islands.

What follows is not a comprehensive survey of Southern California surf sites. Rather, it's a selection of five well-known sites, listed from south to north along the coast (see figure 5.1), that, together, do a good job of illustrating surf science in Southern California.

Black's Beach

Black's Beach is minutes away from urban La Jolla and walking distance from the campus of the University of California at San Diego. But thanks to three-hundred-foot-high sandstone bluffs that run unbroken for several miles, it feels like a piece of wild California, with no city sights or sounds, just the sigh of wind and surf. My favorite times of day are fog-bound mornings and sunny late afternoons. The afternoon brings on the so-called golden hour, when the settling sun lights the bluffs with a bonfire glow. The sunlight shines through the cresting waves, and little rainbows appear in the spume.

Most of the time, the swells at Black's Beach break as lefts. A *left* means that the surfer heads to his left along the breaking face, and a *right* means the opposite. (To someone looking out from the shore, a surfer going left moves to the observer's right, and vice versa.) The lefts at Black's are some of the most legendary in Southern California.

Surf at Black's Beach forms because of the way that approaching swells are bent by Scripps Canyon, a one-mile-long undersea gorge whose head begins just seven hundred feet from the beach. The canyon is a branch of the much larger La Jolla Canyon (figure 5.8). The upper part of La Jolla Canyon follows the Rose Canyon fault (see figure 3.6) for about three miles offshore and then angles west and plunges to more than three thousand feet. Scripps Canyon is five hundred feet deep where it joins La Jolla Canyon a little over a mile offshore. But the continental shelf into which Scripps Canyon cuts is only two hundred feet deep at this point, and approaching swells feel this difference. The part of the swell over the shallow water slows, while the part over the deep canyon keeps going fast. The approaching swells thus bend toward the shallow water to the sides of the canyon, conforming to the bathymetry. The result, shown in figure 5.8, is that wave energy focuses into big surf along the shore just north of the head of Scripps Canyon.

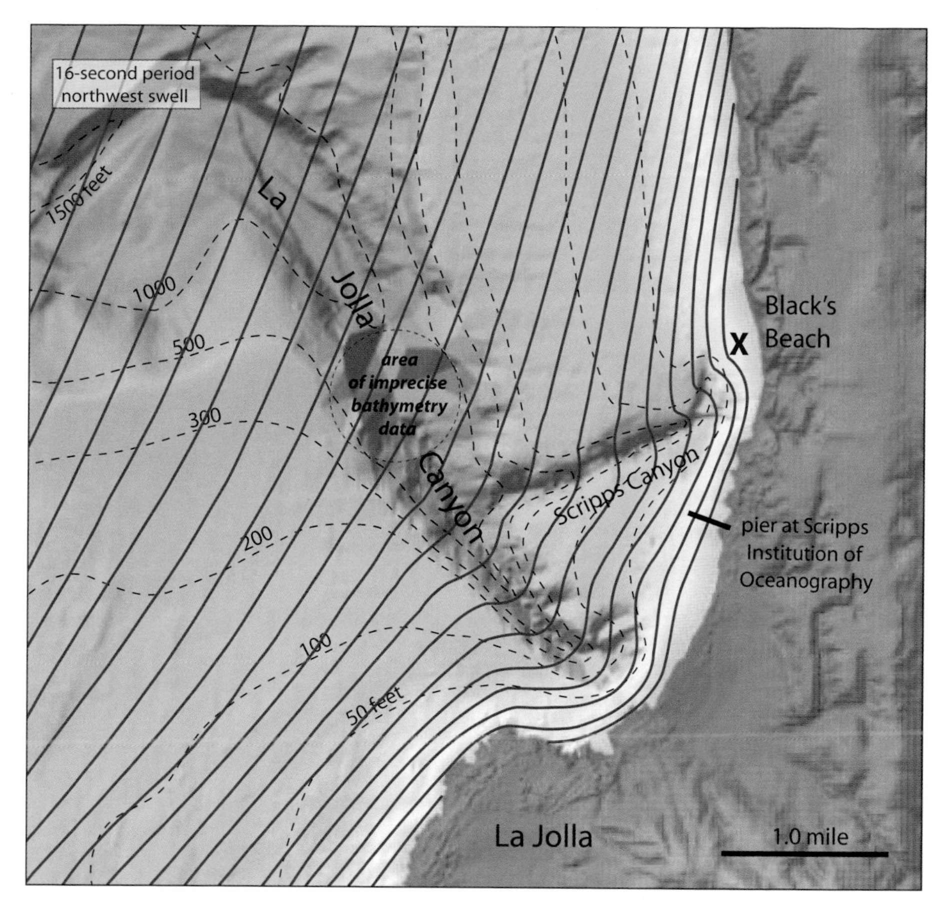

FIGURE 5.8. How refraction of approaching swells over Scripps Canyon creates big surf at Black's Beach. The X marks where the best surfing waves typically break during west-northwest swells. The location of the break will shift north or south, depending on swell direction. Compare this to figure 5.5. (Shaded relief image from NOAA, with labels added.)

Like other mainland surf sites, Black's Beach experiences swell shadows cast by offshore islands, and swell windows through which swells approach unimpeded. Two swell windows produce the best surf at Black's: a northwest window and a southwest window (figure 5.9). These windows are not as sharp-edged as figure 5.9 makes them seem. Unlike light beams, ocean swells bend around island barriers, so even a swell approaching from a shadow direction can deliver good surf if it's powerful enough.

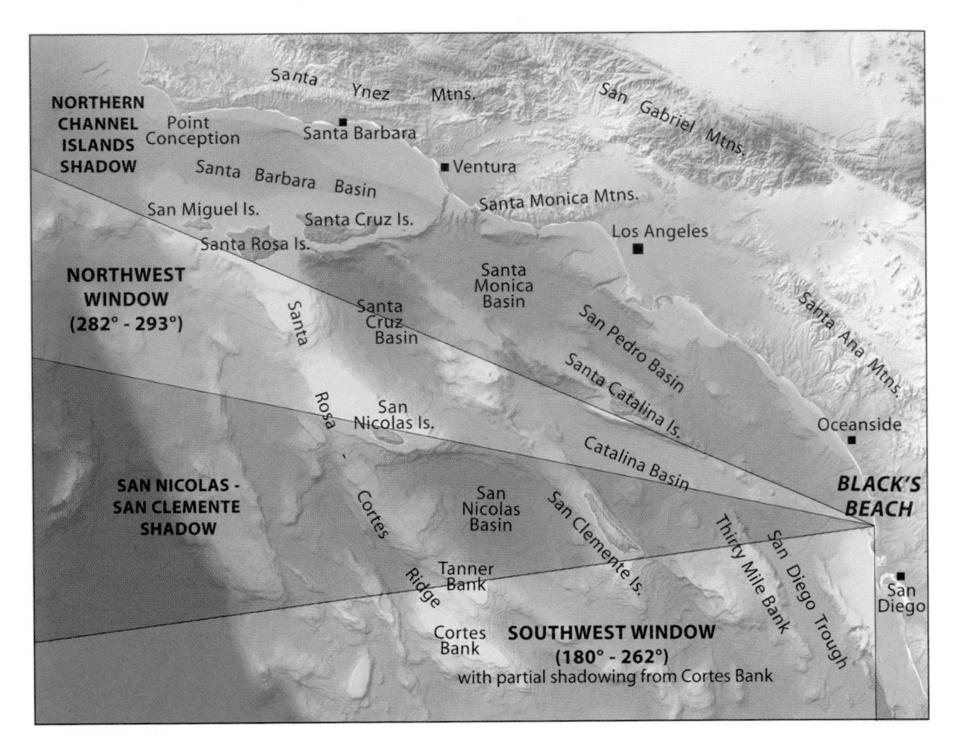

FIGURE 5.9. A swell-window map for Black's Beach. Swell windows occur where open-ocean swells reach the mainland coast through gaps in the offshore islands, whereas swell shadows form where islands block approaching swells. Black's Beach has a narrow northwest window and a wide southwest window. The compass directions, like those on nautical charts, follow a 360-degree circle, with due north = 0°, east = 90°, south = 180°, and west = 270°. (Shaded relief images from NOAA, with labels added.)

Trestles

Trestles, just south of San Clemente, gets its name from an old wooden trestle bridge that once carried the San Diego coastal railroad across San Mateo Creek. In 2012, local governments replaced the deteriorating trestle bridge with a stolid concrete bridge, with *TRESTLES* embossed on it in honor of history, though not design. Walking west under the bridge to the beach, you arrive at the westernmost of five surf spots that make up Trestles. From northwest to southeast, these are Cottons, Uppers, Lowers, Middles, and Church (figure 5.10). The greatest surfing talent usually collects at Lowers, and for pure surfing ballet, it's hard to beat Lowers. Many days of the year there, you'll see turns, spins, and aerials that will drop your jaw. There can be nearly as many telephoto lenses on the

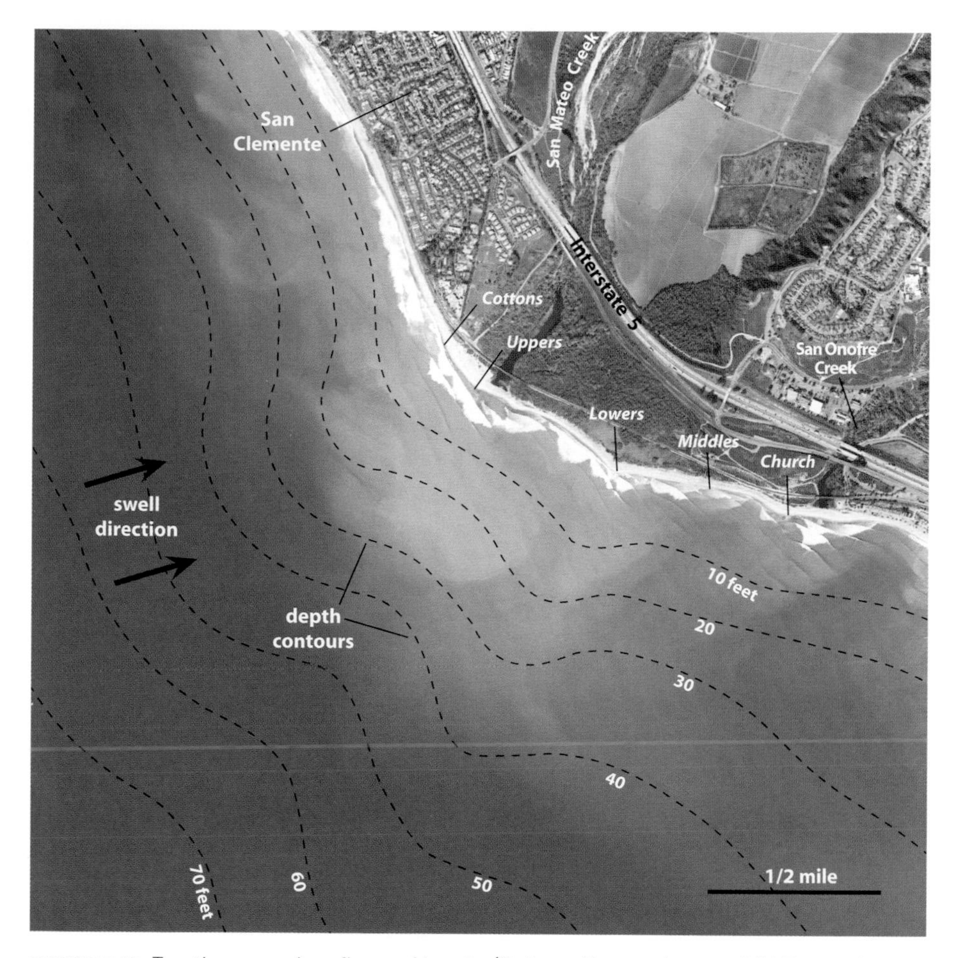

FIGURE 5.10. Trestles comprises five surf breaks (Cottons, Uppers, Lowers, Middles, and Church) located north to south along San Mateo Point. Sand, gravel, and boulders carried by San Mateo and San Onofre creeks during floods have pushed the shoreline seaward to make the bulging point. Approaching swells wrap toward shallow areas formed where the flooding creeks have slewed rock debris into the sea. At the time of this photograph, the swells were approaching from the west-southwest to create the prominent right breaks that you can see along the shore. Depth contours are approximate. (Image ©2014 TerraMetrics; data SIO, NOAA, U.S. Navy, NGA, GEBCO; © 2014 Google; contour lines and labels added by the author.)

beach as surfers in the water, and when the wave is pumping and the talent is in top form, the cameras sound off in quick staccato bursts.

One of my favorite things about surf beaches is the quiet. Yelling or cheering doesn't happen much outside of high-level competitions. Like golf, surfing is a competition against oneself. (Not coincidentally, golf and surfing are the two quietest popular sports.) Even with hundreds of people, Trestles is a quiet place. There are no vehicles aside from occasional lifeguard trucks; the beach is foot- or bike-access only. Although Interstate 5 isn't far, the westerly breeze carries its noise away, so you hear only the cascade of the waves, the scrunch of feet on sand, the clicking of cameras, and the occasional "Ohhh!" when one of the surfers cuts a particularly artful pirouette.

Geographically, Trestles is actually San Mateo Point, a southwest bulge in the coast where two creeks—San Mateo and San Onofre—converge on the ocean. The surf at Trestles comes from the way that swells wrap around shallows formed where the creeks have slewed rocks and gravel into the sea during floods. This makes Trestles an undersea ridge break (see figure 5.4), as opposed to a canyon break like Black's Beach.

Most Southern California creeks are bone dry much of the year. But during heavy rains, they can become raging torrents thick with mud and bouncing boulders. You may never see a Southern California creek in a big flood, but that says more about the limits of a human lifetime than about geologic reality. After the Earth warmed out of the last ice age some twenty thousand years ago, global sea level rose nearly four hundred feet before stabilizing close to its present level about seven thousand years ago. The world's creeks and rivers have thus been discharging into the ocean at close to the same position, in relation to sea level, for seven millennia. A truly massive flood may sweep down San Mateo or San Onofre creeks just a few times per century—but over seven millennia that translates into several hundred big floods, each of which delivers a load of sand, gravel, and boulders to the shoreline. Collectively, these floods have pushed the shoreline seaward nearly a half-mile to make San Mateo Point. The flooding creeks didn't always dump their loads in the same place; their outlets shifted north or south. The result is a series of bouldery bulges that stick out from San Mateo Point to make the surf breaks at Trestles. Approaching swells slow down more over the shallow bulges than in the deeper water to the sides, so that the waves wrap into the bulges (figure 5.10). At Lowers, a bouldery bulge extends nearly a mile offshore, wrapping southwest-approaching swells to make classic A-frame breaks with double barrels that peel off both left and right. It wasn't always that way. Before the wet El Niño winter of 1982–83, Lowers was mostly a right-breaking wave. A fresh load of El Niño gravel changed that, so that now Lowers often sports a good left wave along with its classic right. After a big rainstorm, surfers at Trestles may end up dodging logs, branches, drowned jackrabbits, and assorted urban trash. The surf breaks at Trestles are geologic works in progress, subject to revision with every flood.

Redondo Breakwater

On the wide, nearly flat Santa Monica Shelf west of Los Angeles, the water can be less than three hundred feet deep even ten miles offshore. But Redondo Canyon cuts a deep defile across this shallow expanse, reaching practically to the breakers at Redondo Beach (figure 5.11a). The topographic shift from the near-flat shelf to the raw-edged canyon is one of the most dramatic in California; if it were on land, it would be a bit like coming upon a mini–Grand Canyon in the middle of the Los Angeles Basin. What's the canyon doing there? Figure 5.11a gives a partial answer. Canyons, on land or at sea, often follow faults because erosion works faster on the weak and pulverized rock along faults. Just as the upper part of La Jolla Canyon near Black's Beach lines up with the Rose Canyon fault (see figure 3.6), so Redondo Canyon likewise lines up with the Redondo Canyon fault, which splays off the Palos Verdes fault north of the Palos Verdes Peninsula. (There's actually more to the story of how submarine canyons form; see text box in chapter 6 for a fuller explanation.)

Redondo Canyon aims about twenty compass degrees south of due west, and when a long-period swell approaches from that direction, the refraction effects are similar to Scripps Canyon near Black's Beach. The approaching swells slow in the shallow water over the Santa Monica Shelf but continue fast in the deep water over the canyon, bending away from the canyon to focus their energy toward the breakwater outside the Redondo Beach and King Harbor marinas (figure 5.11b). Redondo doesn't fire off big surf very often; it takes just the right direction and size of swell. But when it does, it can be awesome. On December 5, 2007—known locally as Big Wednesday—a seventeen-second-period swell unloaded titanic surf all along the coast. The swells lined up near perfectly with the axis of Redondo Canyon, and as they bent away from the canyon and hit the breakwater, they exploded in wet fireworks fifty

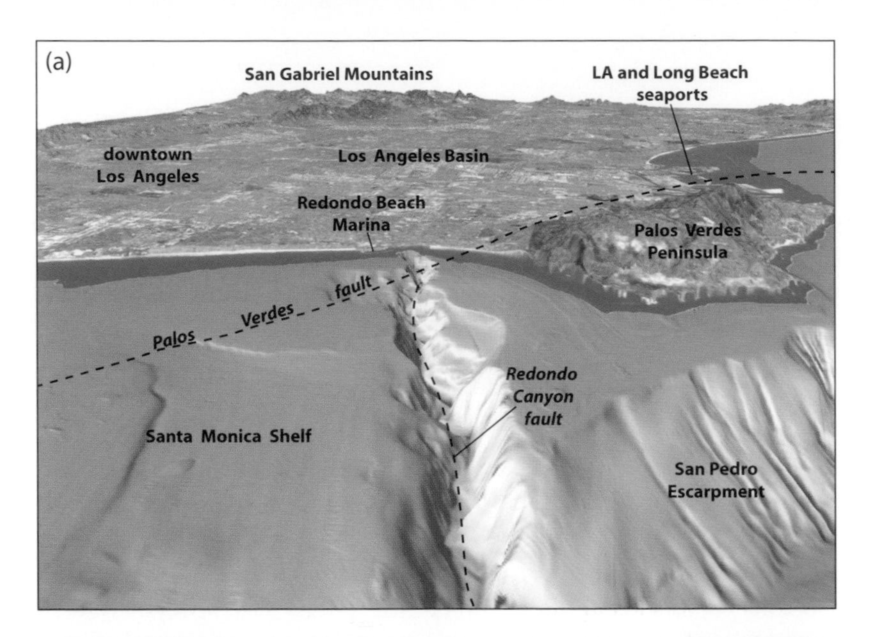

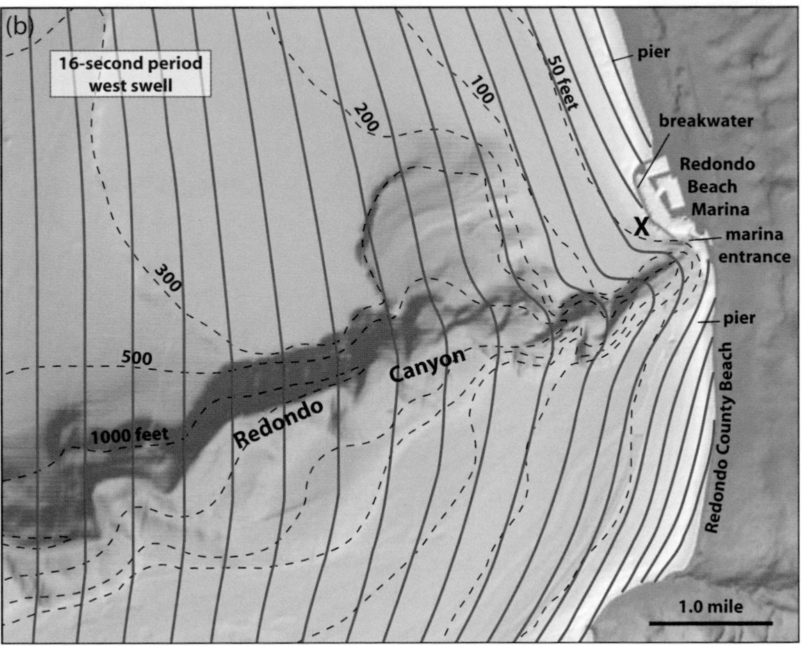

FIGURE 5.11. Nine-mile-long Redondo Canyon notches into the Santa Monica Shelf west of Redondo Beach. (a) Perspective view of the canyon derived from multibeam bathymetry, looking east toward Los Angeles from about twelve miles offshore. The image exaggerates the topography; the canyon walls are not nearly so steep. The canyon is three hundred feet deep just a half-mile from the beach and descends to two thousand feet at its mouth, where it exits the San Pedro Escarpment. (Image from U.S. Geological Survey Pacific Coastal and Marine Science Center, with labels added.) (b) A refraction diagram showing how Redondo Canyon bends long-period west swells, focusing their energy toward the breakwater outside the Redondo Beach and King Harbor marinas. (Shaded relief image from NOAA, with labels added.)

feet high. Long left tubes as high as twenty feet peeled northward from the breakwater, slingshotting surfers onto some of the wildest rides of their lives.

Entering the water during big surf at Redondo Breakwater is an adventure. To get to the break, many surfers opt to bypass the exhausting whitewater paddle out from the beach north of the breakwater, and instead walk several hundred yards seaward along the north portion of the breakwater. (The north portion, because of its curve, is somewhat protected from swells bent northward from Redondo Canyon; figure 5.11b.) Timing their leaps off the breakwater, the surfers land just behind a passing wave and then paddle like hell, since mistiming the entry risks being meat-axed against the breakwater by the next wave. I walked out to the entry point on a moderate day (head-high tubes) and was scared out of my wits. It didn't help that a bloody half of a dolphin had recently washed up on the beach. Great white sharks congregate in Redondo Canyon, where the depth close to shore favors their preferred mode of attack: vertical from below, hitting their prey—seals, sea lions, and dolphins mostly—so hard that sometimes the whole shark leaves the water, tail still pumping. While scuba diving, I've looked up at wetsuited limbs protruding from surfboards. My human brain knows they aren't seals, but the resemblance is clear. Can a fish brain backed by a thousand pounds of hunger be more discriminating?

Malibu Point

If there is one place synonymous with surf culture in Southern California, it's Malibu. Some call it the birthplace of surfing, which is going too far (ancient Polynesians invented the sport). But it's fair to say that Malibu is to surfing what Ted Williams is to baseball—a name that will endure and always come up in discussions about the "greatest." More than seventy-five surfing films have featured Malibu, and in 2010, it was designated as the first official World Surfing Reserve.

Malibu is one of Southern California's classic point breaks, meaning that the waves wrap around a point of land that bulges into the sea. Malibu, like Trestles, lies at the mouth of a creek. Floods down Malibu Creek over past millennia have slewed gravel and boulders into the ocean, pushing the shoreline several thousand feet seaward to create Malibu Point, around which approaching waves bend and focus their energy. Figure 5.12 shows the break during a big southwest swell. You can see how the waves bend fully ninety degrees as they refract around

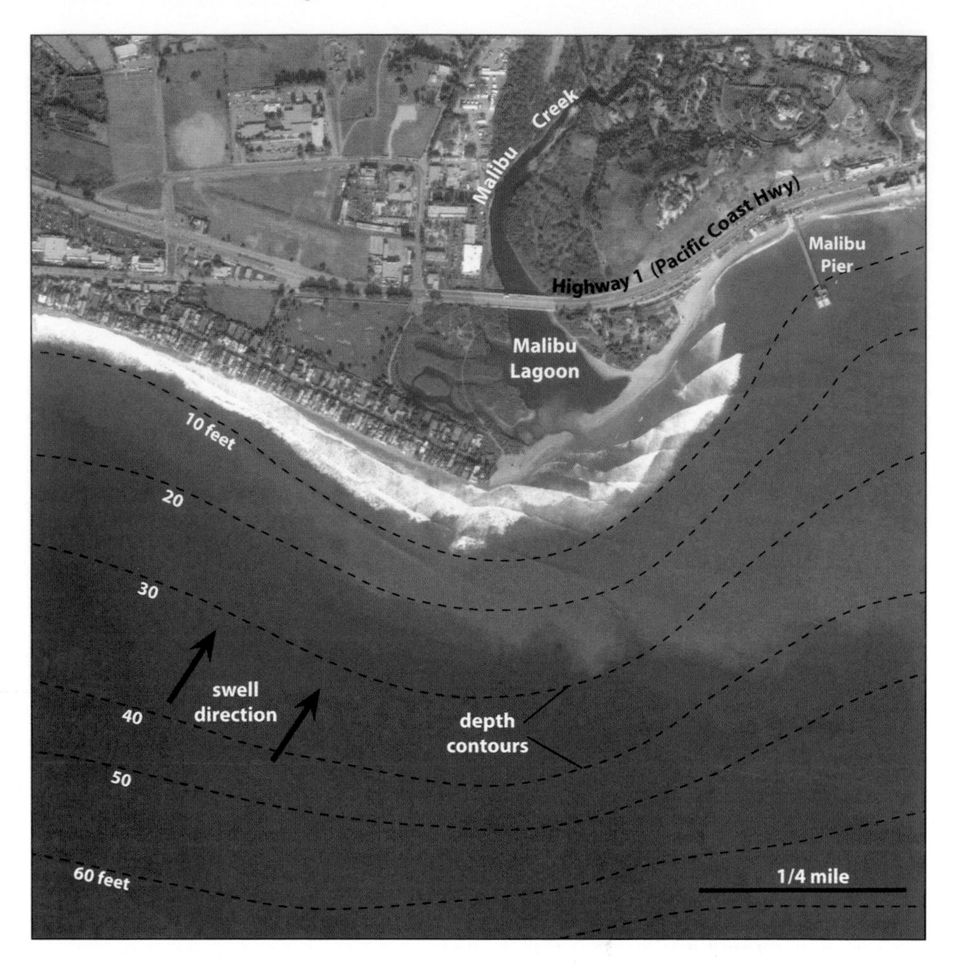

FIGURE 5.12. Malibu Point, the first official World Surfing Reserve. Floods down Malibu Creek have pushed the shoreline seaward to make the bulging point. At the time of the photograph, the swells were approaching from the south–southwest and wrapping to the point, focusing their energy to create the prominent right breaks visible in the surf zone. Notice how refraction has turned the waves almost ninety degrees from their original approach. (© 2014 Google; data CSUMB SFML, CA OPC; image © 2014 TerraMetrics; contour lines and labels added by the author.)

the point. On days like that, surfers can catch rights that may carry them for more than a quarter-mile.

Malibu enthusiasts speak of its "perfect wave"—meaning its long, peeling, classic rights—but perfection doesn't come that often. Part of the reason is Malibu's location, on a south-facing coastline east of Point Dume and the Northern Channel Islands. These barriers block Malibu

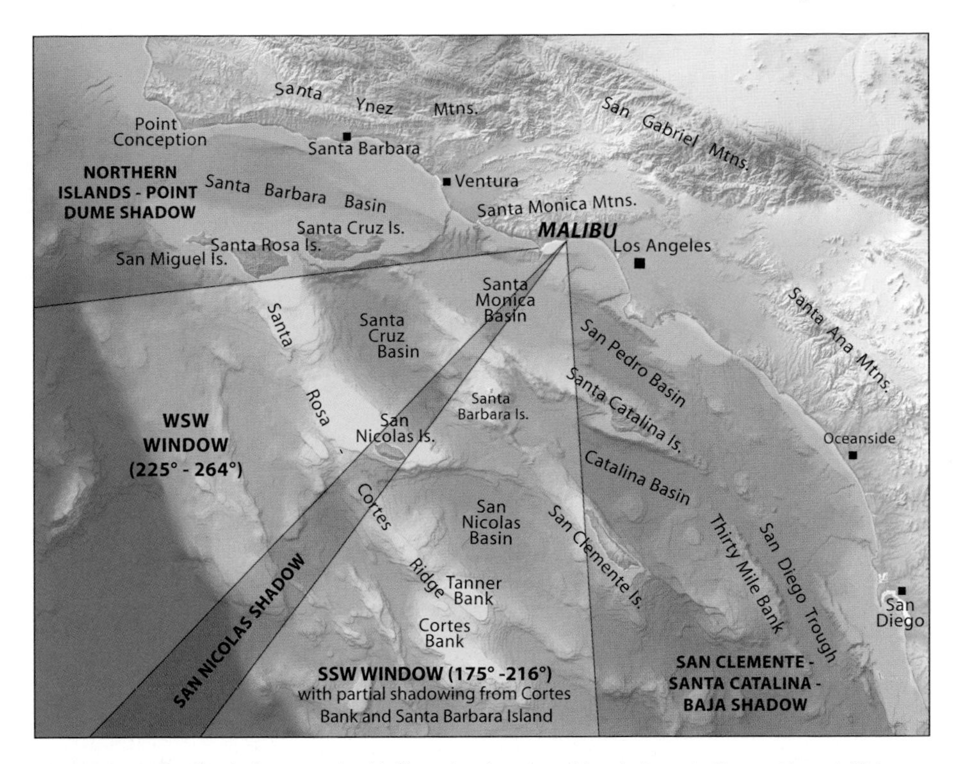

FIGURE 5.13. Swell-window map for Malibu, showing two wide windows to the southwest. This southwest exposure makes Malibu one of Southern California's classic summer surf sites. The compass directions, like those on nautical charts, follow a 360-degree circle, with due north = 0°, east = 90°, south = 180°, and west = 270°. (Shaded relief images from NOAA, with labels added.)

from the west and northwest swells that dominate the California coast during late fall, winter, and early spring. Summer is Malibu's prime season because that's when big south swells arrive. Malibu Point is perfectly positioned to receive southwest and south swells through gaps between the offshore islands (figure 5.13). You may recall from earlier that southwest swells are California's farthest traveled, arriving here from the storm-tossed South Pacific (figure 5.2; see also the text box earlier in the chapter). Malibu's surf is Southern Hemisphere storm energy or tropical hurricane energy delivered to Southern California's beaches.

Rincon Point

Rincon is a Spanish word for an angular bend in a coastline, mountain, or river. There are plenty of rincons along the California coast, but

FIGURE 5.14. Rincon Point along the Ventura–Santa Barbara county line. The view looks west in the direction from which the swells are approaching. As westerly swells refract around the point, they form right breaks that can give rides of more than a quarter-mile. Notice how refraction has turned the waves almost ninety degrees from their original approach by the time they wrap around the point. (Photograph by Woody Woodworth.)

none more graceful than the one that angles seaward fourteen miles west of Ventura.

Like Malibu and Trestles, Rincon Point is a river-mouth break. Floods down Rincon Creek have pushed the shoreline seaward and spread bouldery debris well beyond, forming shallows around which approaching swells wrap and focus their energy. Both Rincon and Malibu are legendary for their long, well-formed rights. But the similarity ends there. Malibu, the town, is an upscale extension of Los Angeles, with frou-frou shops and resident movie stars. Rincon Point has no town, no shops, no snack stands, and no beach kitsch—just a free parking lot next to a row of quiet beach houses. Malibu and Rincon also have opposite seasons. Malibu is blocked from most winter swells but open to summer swells. Rincon is the opposite. The Northern Channel Islands block Rincon from most south and southwest swells in summer, but it lies wide open to westerly winter swells cruising down the thirty-mile-wide Santa Barbara Channel. As these swells meet Rincon Point, their refraction is dramatic (figure 5.14). Wrapping into the point, the swells peel off as long, barreling rights. Surfers may stand up in Santa Barbara County and end their rides, sometimes more than a quarter-mile later, in Ventura County, since the county line bisects the point.

6

Beaches and Coastal Bluffs

The coast is the scene of never-ending struggle among natural and human forces. The ocean batters the shore, pulling sand away from beaches and undermining bluffs and houses, while an ever-growing population presses in from the landward side toward the water's edge.

—California Coastal Commission

Point Fermin sits at the southern tip of the Palos Verdes Peninsula, less than a mile from the seaport of Los Angeles. The hills of San Pedro rise steeply, with terraced streets densely lined with apartment houses and bungalows. But there aren't any streets or houses on the ten acres of Point Fermin closest to the ocean. Instead, there's Sunken City: the slumped and cracked remains of a community destroyed as the land broke up and slid downhill toward the sea.

A high fence posted with No Trespassing signs blockades Sunken City. But people go in anyway, through breaks in the fence. Strolling through the ruins is like wandering the set of a post-apocalyptic movie. Roads sail off into open space. Blocky, angular sections of streets jut at angles from the ground. Chasms cleave foundations, and broken pipes and sewer lines poke from fractured cliffs. Graffiti artists have covered the concrete surfaces with swirling designs. On my meanderings, I met several dog-walkers, a model with her photographer, and a teenage couple emerging from a secret hideaway among the ruins, carrying a blanket and looking rather tousled.

The land slumps in a set of irregular steps to a rocky beach about a hundred feet below. Each step is an individual slump block (figure 6.1). The exposed faces of the slump blocks reveal the geologic reason for the destruction of Sunken City. The bedrock consists of sedimentary layers that incline toward the ocean like a tilted deck of cards. Such an

FIGURE 6.1. Sunken City, the remains of a neighborhood destroyed by the slow collapse of coastal bluffs at Point Fermin in the city of San Pedro. Movements began here in 1929 and have continued periodically since. Low cliffs called *scarps* mark where sections of bedrock have detached and slid away. Chronic bluff erosion problems like this plague much of Southern California's coast. (Photograph by Bruce Perry, California State University, Long Beach.)

arrangement results in what geologists call *day-lighting*, meaning that the ends of tilted rock layers see daylight along bluffs and slopes. It's an unstable situation (picture cards sliding as you tilt a deck), but not necessarily disastrous. Two other factors, however, have conspired to make the situation at Sunken City profoundly unstable. The first is the undercutting of the coastal bluffs by ocean waves. The narrow, rocky beach below the bluffs offers scant protection from wave attack, especially at high tide. The second factor is a thin bed of bentonite sandwiched within the stack of rock layers. Bentonite is a type of clay that forms when layers of volcanic ash are exposed to water and air. Bentonite becomes unbelievably slippery when wet and is the scourge of dirt roads in many parts of the West. (In backcountry Wyoming, I've seen four-wheel-drive trucks hopelessly mired in bentonite. "Four-wheel-drive just gets you stuck worse," one rancher told me.) When the bentonite bed at Point Fermin gets wet, the rock layers above it begin to slide toward the sea. The sliding began in 1929, forcing abandonment of the houses in the area, and has continued on and off ever since.

In Southern California, human settlement and geologic circumstances have conspired to produce some of the nation's worst coastal erosion problems. The region is sandwiched within an active plate boundary where fault movements fracture the rocks, push them up into steep slopes, and shake those slopes periodically with earthquakes. Waves continually chew out the bases of coastal bluffs. Much of the bedrock along the coast is made up of weak sedimentary layers prone to failure, especially if they happen to tilt toward the ocean. Wetness is a major culprit in coastal erosion because water lubricates and weakens rock. Urban settlement increases wetness; think of all the water that goes onto lawns and gardens, along with leakage from underground pipes. Moreover, human development has shrunk the region's beaches (for reasons I'll explain later in this chapter) so that coastal bluffs are now less buffered from wave attack.

The wettest conditions in Southern California usually come in winter, when storms spawned in the Gulf of Alaska sweep southward down the coast. The rains leave the coastal bluffs soaked and weakened, and the storm waves strip sand from the beaches and gnaw at the unprotected bluffs. Inevitably, portions of the bluffs collapse, taking away backyards, lawns, swimming pools, and even parts of houses. Disaster footage highlights the nightly news, insurance claims are filed, and then—more often than not—rebuilding commences. The battle to restore damaged oceanfront property involves a multipronged counterattack employing concrete, rebar, caissons, steel mesh, fill dirt, seawalls, and drainpipes, as if the root of the problem were engineering rather than location.

If you've lost property to bluff erosion, it may be hard to accept that the retreat of California's coastal bluffs is natural and normal. As the latest ice age ended about twenty thousand years ago, the ice sheets that once cloaked much of North America and Eurasia with up to two miles of ice began to melt, and world sea level began to rise. The sea rose about four to five feet per century from fifteen thousand to seven thousand years ago, for a total rise of close to four hundred feet. Then, about seven thousand years ago, the rate of rise slowed, and in recent millennia, the sea has crept upward, on average, less than two inches per century (figure 6.2). This means that for the past seven thousand

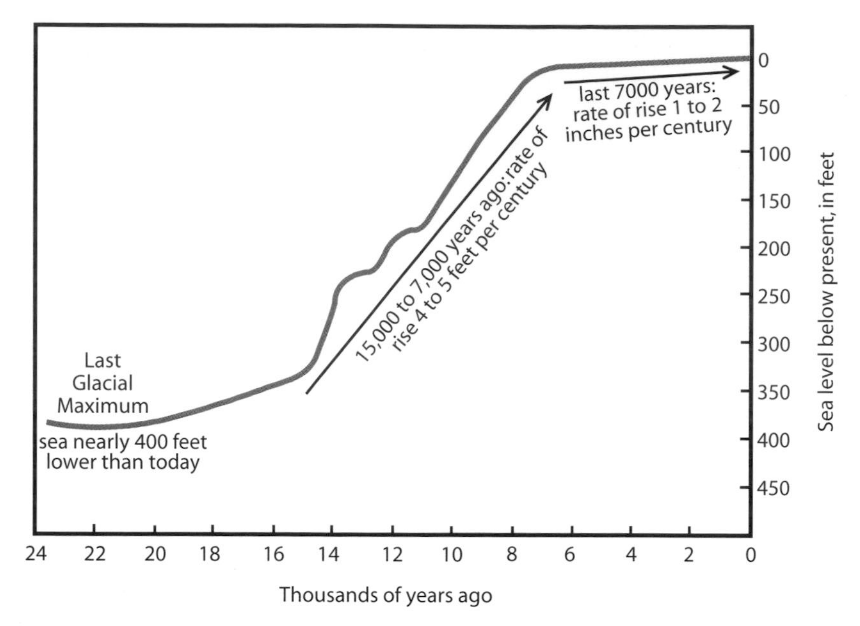

FIGURE 6.2. Sea-level rise since the Last Glacial Maximum twenty thousand years ago. Notice that during the most rapid period of postglacial rise, between about fifteen and seven thousand years ago, sea level rose four to five feet per century. About seven thousand years ago, the rate slowed to one to two inches per century. This means that for seven thousand years, ocean waves have been attacking the coastlines of the world at about the same level, causing coastlines worldwide to retreat by hundreds to thousands of feet. The rate of sea-level rise today is accelerating with global warming. Global sea level has risen seven inches in the past hundred years, and most forecasts project between two and six feet of rise by 2100—in other words, returning roughly to the rate of rise that took place between fifteen thousand and seven thousand years ago. We'll explore the implications of past and future sea-level changes in chapter 7. (Based on Peltier and Fairbanks 2006.)

years, the sea has been chopping away at the land at nearly the same level worldwide, pushing coastal bluffs back by hundreds of feet all over the planet. The coastal bluffs we see today represent a snapshot moment in this long history of retreat—one that began before the oldest pyramids arose from the sands of Egypt (figure 6.3).

Rates of bluff retreat vary widely along the coast because of shifts in the strength of rock and exposure to prevailing waves. Where granite or tough metamorphic rocks line the coast (as they commonly do north of Point Conception), long stretches of coastal bluff have barely budged in historical time. But the soft sedimentary rocks that line much of the Southern California Bight erode, on average, a few inches per year, and even as much as a foot per year in some places. Those are long-term averages; erosion rates are highly episodic. A section of bluff may sit stable for decades, promoting complacency—"My bluff hasn't moved in thirty years, it's stable"—and then suddenly several feet will plummet down slope. Coastal bluffs fail unpredictably and without warning (figure 6.4), but if you want to put money on when it's most likely, pick a wet winter day or night, during high tide, when storm waves are battering the base of the bluffs.

COASTAL EROSION: RETREAT OR HOLD THE LINE?

"I get it that bluff erosion is natural," you may say, "but we can't just sit back and let waves wreck valuable oceanfront property." Actually, we could. Many communities will eventually have to pull back from the coast, and in some places, they already have. At Surfer's Point in Ventura, construction crews have torn out old bike paths and parking lots that waves have partly destroyed and relocated them inland, buying perhaps another fifty years. Likewise, at Goleta Beach County Park near Santa Barbara, work is under way to move bike paths, parking lots, and buried utility lines back from the wave-chewed bluffs. This approach—called *managed retreat* or *planned retreat*—represents a profound shift in thinking for California, a state that, until recently, crammed its roads, homes, and businesses right up to the ocean's edge and threw up miles of seawalls and revetments (piles of large rocks) against the waves. Many coastal communities are currently developing plans for managed retreat, prompted, in part, by forecasts of sea-level rise in upcoming decades. Recall from figure 6.2 that the rate of sea rise since seven thousand years ago has averaged less than two inches per century. But in the past hundred years, global sea level has risen seven inches, and recent projections by the National Research Council forecast two to six feet of rise by the year 2100.

Managed retreat can be most easily managed on public land. It's more problematic for private property. But here, too, there are options. If bluff-top property owners have acreage, they can move their houses back from the bluff edge and thus buy themselves some time. In some instances, local governments can buy bluff-top properties, remove the structures, and let erosion take its course. For example, in 2002 the city of Pacifica, south of San Francisco, purchased a row of bluff-top homes that were about to be taken by the sea and demolished them. Large-scale buyouts of private coastal property have yet to occur in Southern California, but time and sea rise will eventually weigh in. If erosion degrades someone's property to the point that it becomes legally condemned, then retreat—managed or not—will have occurred.

FIGURE 6.3. Coastal bluffs make up more than half the coastline of the Southern California Bight. The rocks are mostly weak sedimentary layers that erode two to four inches per year on average, although the erosion is highly episodic. The view looks north along the bluffs above Black's Beach near La Jolla (site of the famous surfing beach; see figure 5.8). Assuming an erosion rate of three inches per year (typical for this section of the San Diego coast), the dashed lines show roughly where the bluff edge probably stood at various times in human history. (Photograph by the author.)

Jesus of Nazareth
~2000 years ago
Columbus crosses Atlantic
~500 years ago

FIGURE 6.4. Bluff collapse caught in action at Torrey Pines State Beach near La Jolla. Coastal bluffs erode episodically, often by instantaneous collapses like this. Comparison of the texture and composition of sand on the beach with that of sand derived from bluffs and from rivers indicates that bluff erosion—where not slowed by seawalls—provides as much as half of the sand on some Southern California beaches. (Photograph by Herb Knufken.)

While managed retreat may be the best long-term option for coastal communities, in the short term—meaning the next few decades—it's not unrealistic to try to hold the line. We can't completely stop coastal bluffs from eroding, but in many places we can stabilize them for a while. There are two main ways to do this: *hard stabilization* and *beach replenishment.* Hard stabilization means armoring bluffs with concrete seawalls or revetments. Beach replenishment means importing sand to widen beaches, thus buffering bluffs from wave attack. I'll come back to beach replenishment later in the chapter, but here I'll say a few things about hard stabilization.

Bluff-top property owners get warm and fuzzy near concrete—their main weapon in their battle to protect the bluffs. "I *love* my new seawall!" one fellow enthused to me as we stood on the beach below his property in Solana Beach. The wall stood about twenty-five feet high, rising nearly halfway up the bluff face, and stretched sixty feet from end to end. He ran his hands across the concrete. "It's rather nice-looking, isn't it?" I had to agree. The wall was skillfully contoured and colored to mimic the look of the natural bluff. "This buys me at least twenty years of stability, by which time my kids will probably ship me off to an old-folks home and sell the house," he laughed. We walked away from the seawall so we could look up to his house on the bluff top. The ends of the wall, traced upward, lined up with the edges of his property. Stretch a measuring tape along a Southern California seawall and you'll very often find that it's fifty, sixty, or eighty feet long, or some additive combination. That's because those are typical widths of bluff-top lots. Property owners, not public monies, pay for seawalls that protect private property, so the edges of seawalls usually line up with property boundaries (figure 6.5). Sometimes, bluff-top neighbors will pool funds to get an extra-long seawall that spans several properties, thus saving some costs. I could see that waves had excavated several deep caves into the bluff on either side of the seawall. "I tried to get my neighbors in on the project," the fellow said, "but they couldn't afford it. So I had to go ahead alone."

Had to go ahead? "Oh yeah, no question about it," he said. "The bluff below my place was in such bad shape that I could have lost my house. That's the only way you can get a CCC permit these days. You have to show that your structure is imminently threatened by bluff collapse. I was holding my breath that the bluff wouldn't go before I got my permit."

The "CCC" to which he referred is the California Coastal Commission—the entity that controls construction in much of California's coastal zone. As part of its mission to "protect, conserve, restore, and enhance environmental and human-based resources of the California coast," the CCC grants or denies permits to build seawalls. The CCC doesn't support seawalls—in fact, it is clear about their adverse effects: "The loss of beaches due to armoring [which includes seawalls] and sea level rise results in immense negative impacts, including loss of recreational value, tourism, marine mammal haul-out area, sandy beach habitat, and buffering capacity against future bluff erosion." But the CCC's mission is to protect both the environmental and the human-based resources of the coast, which puts it in a tough position. On the one hand, seawalls lead to shrinking beaches (for reasons I'll explain shortly); on the other hand, seawalls are sometimes necessary to protect property. (Seawalls represent a singular private-versus-public property conundrum because most are built on public beaches to protect private property on the adjacent bluff top.) The CCC walks the line between

FIGURE 6.5. Seawalls slow bluff erosion and protect bluff-top properties, but they also cut off an important source of new beach sand (see figure 6.4). Throughout much of California, the California Coastal Commission (CCC) regulates construction on bluff faces. The CCC requires, among other things, that new seawalls are built to mimic the natural rock of the bluff face, as this example from Solana Beach near San Diego shows. Property owners pay for seawalls, which is why the ends of seawalls usually line up with property boundaries on the bluff top. (Photograph by the author.)

these competing interests with a two-pronged approach. First, it typically grants seawall construction permits only when bluff collapse imminently threatens the primary structure on the property. A certified engineering geologist makes this judgment, and the threat must be to the primary structure; no one gets a seawall permit just to protect his backyard or barbecue pit. Second, the CCC sets substantial mitigation fees to compensate for the loss of beach sand and beach space caused by the seawall. These fees go toward beach restoration and improving public beach access.

"I'm sure your CCC permits and fees set you back a fair sum," I told the fellow on the beach. "How about the wall itself; what did that cost you?" His answer took me aback: more than $300,000. Not quite believing that a concrete wall could cost more than a nice family home in some areas of the nation, I called the owner of a seawall construction company, who gave me a concise primer on modern seawall costs. The work can happen on only a few days each month, during the extra-low tides that accompany the new and full moons. The wall needs to stand on bedrock, so workers need to dig down through as much as fifteen feet of sand at the base of the bluff to reach the wave-cut bedrock below. The wall must conform to the shape of the bluff and thus needs to be constructed on site, not prefabricated. Deep holes need to be drilled into the bluff to fasten the wall firmly to it. Drain holes need to be installed; otherwise, groundwater will accumulate behind the wall, weakening the rock and putting outward pressure on the wall. If open fractures or caves exist in the bluff—as they often do in areas threatened by imminent collapse—these need to be sealed before the wall can go up. One of the most specialized parts of the construction process involves making the molds for casting the concrete. The molds are built so that the concrete will mimic the natural lines and roughness of the bluff, and the concrete is dyed to match the rock (figure 6.5). Ever since the 1990s, the CCC has mandated that new seawalls have a natural look to minimize their aesthetic impact. I, for one, am grateful. Seawalls built before the naturalistic requirement are about as attractive as the walls of a maximum-security prison, and many still blight California's beaches.

BLUFFS TO BEACHES—WHAT'S THE CONNECTION?

The topic of bluff erosion is linked inextricably to the health of California's beaches. A wide beach provides the best buffer against wave erosion, with added benefits for public use, tourism, and safety (fewer bluff collapses to kill people). Beaches pump vast amounts of money into the economies of California's coastal cities, mostly from tourism and property taxes on valuable oceanfront real estate. The California Department of Boating and Waterways estimates that the state's beaches generate more than $15 billion in tax revenue annually.

California beaches naturally change size—often dramatically—between summer and winter. Summer produces few storms in the North Pacific, so summer waves are usually small. These small summer waves push sand from the surf zone toward shore, building up wide beaches. But winter brings storms and larger waves, which sweep sand off the beach and carry it out to the surf zone. The following summer, smaller waves push the sand back. The result is a predictable seasonal change in the size of the

FIGURE 6.6. California's beaches change seasonally with the size of waves. (a) Winter storms bring large waves that sweep sand off the beach and out into the surf zone, exposing wave-rounded rocks that are usually buried by sand in summer. (b) Summer brings smaller waves that push sand from the surf zone back up onto the beach, covering the rocks and building a wide, sandy beach. (Both photographs were taken by the author at the same location in Solana Beach, six months apart.)

beach (figure 6.6). But superimposed on this seasonal cycle is a troubling longer-term trend. Many of California's beaches are experiencing net shrinkage. The blame can be traced directly to human construction—specifically to river dams, seawalls, and harbor breakwaters and jetties.

Matilija Dam lies deep in the Santa Ynez Mountains along Matilija Creek, a tributary of the Ventura River. The Santa Ynez Mountains, crushed as they are in the Big Squeeze (chapter 3), are arching upward by as much as a half-inch per year in places, and it's all the rivers can do to keep up. The rivers and creeks have cut downward apace with the rising mountains but have had no time to widen their valleys, so they lie trapped in narrow canyons of their own making. When the final bucket of concrete was poured for Matilija Dam in 1948, the dam's limited life span was well understood. Even the Army Corps of Engineers—one of the most pro-dam-building entities in the nation—concluded that the dam would fill with sediment within a few decades and not provide enough benefit in water storage and flood control to justify its costs. Yet the dam went up, and there it sits today—abandoned and crumbling, its reservoir nearly filled in with sediment, but still catching every grain of sand that might otherwise move downstream to widen Ventura's beaches (figure 6.7a).

Enough sand and silt now plug Matilija Reservoir to fill three football stadiums to their brims (roughly six million cubic yards). The same thing is happening behind hundreds of other dams on coastal rivers throughout California. Dams shrink beaches in two ways: by trapping sand that the rivers would normally carry to the coast, and by dampening floods. In Southern California, where rivers are meager and usually intermittent, floods—typically during winter rainstorms—are the main way that sand moves downstream to the beach. Further cutting into the supply of river sand for beaches is widespread mining of sand and gravel from riverbeds for construction, and lining of riverbanks with concrete to prevent lateral erosion in urban areas. (Miles of the Los Angeles, San Gabriel, and Santa Ana "rivers" are now concrete-lined channels.) These developments mean that the major rivers of Southern California (from north to south: the Santa Maria, Santa Ynez, Ventura, Santa Clara, Los Angeles, San Gabriel, Santa Ana, Santa Margarita, San Luis Rey, and San Diego rivers) deliver from one-half to one-fifth of the beach sand that they once did.

Attrition by river dams accounts for just part of the sand deficit of Southern California's beaches. Many beaches gain a large fraction of their sand from natural bluff erosion (figure 6.4). Chunks of bluff fall to the beach, and the waves gradually break those chunks down to form new sand. So, where seawalls or revetments slow bluff erosion, the adjoining beach shrinks. It's straightforward to estimate—at least roughly—the bluff-erosion component of beach sand. If we know the average annual rate of natural bluff retreat (which we can get from historical records, such as old maps showing where bluff edges used to be),

FIGURE 6.7. Examples of two Southern California dams whose reservoirs are now mostly or entirely silted in. (a) Matilija Dam on Matilija Creek, a tributary of the Ventura River in the eastern Santa Ynez Mountains. Originally 198 feet high, the dam was notched 30 feet in 1965 to reduce pressure on inferior concrete. In 1977, the notch was widened to the position seen here. (Photograph by the author.) (b) Rindge Dam on Malibu Creek, about two miles upstream of Malibu. Sediment that flowed down Malibu Creek during prehistoric floods built Malibu Point—one of California's most famous surf sites (see figure 5.12). The dam, built by a rancher in 1926, had backfilled completely by the 1950s and continues to retain vast amounts of sand that might otherwise widen Malibu's beaches. (Photograph by the author.)

we can calculate the volume of bluff rock that would normally fall onto the beach each year to potentially make new sand.* For instance, about thirty-five miles of bluffs, averaging about fifty feet high, line the San Diego County coast from Camp Pendleton to La Jolla. Assuming an average natural erosion rate of two to four inches per year, this would yield between 57,000 and 114,000 cubic yards of new beach sand each year if not slowed by seawalls. That's 20–40 percent of the 275,000 cubic yards of sand that move along San Diego's beaches each year by longshore drift. In other words, sand from bluff erosion forms a substantial portion of the sand budget of these beaches. Research confirms this. Recently two independent research groups at the University of California at San Diego concluded that bluff erosion supplies roughly half of the sand on many of northern San Diego's beaches.

Jetties and breakwaters interfere with beaches too, although in a different way. Whereas river dams and seawalls cut down on the supply of *new* sand that reaches the beach, jetties and breakwaters interfere with the transfer of *existing* sand along the beach. Vast amounts of sand move along the beach by longshore drift—the process whereby waves nudge sand down the coast whenever they approach the beach at an angle. Ocean swells from the west and northwest dominate the Southern California Bight. These swells usually break against our beaches at an angle from the north, so that the breakers sweep from north to south and steadily push the sand south. On average, Southern California beaches experience net southward longshore drift of 250,000 cubic yards of sand per year. (For perspective, that's about 25,000 dump truck loads, which equals a line of bumper-to-bumper dump trucks stretching for 120 miles.) Jetties and breakwaters (built to protect harbor mouths or to keep lagoon inlets open) trap this south-moving sand, causing beaches on the down-drift (south) sides of the structures to shrink, while the beaches on the up-drift (north) sides grow. The accumulating sand on the up-drift sides eventually plugs harbor mouths if it isn't dredged regularly (figure 6.8). Indeed, the volume of sand that must be dredged to keep harbor mouths open gives us our best estimates of how much sand moves along Southern California beaches by longshore drift. Two hundred fifty thousand cubic yards per year is the average, but the amount varies widely from place to place as a result of changes in wave energy and sand availability

* Not all eroded bluff rock becomes beach sand; some types of bluff rock erode into blocks and boulders instead of sand. But the relatively soft sandstones that form the bluffs along much of the Southern California Bight break down readily into new sand once pieces of the bluff fall to the beach and are bashed apart by waves.

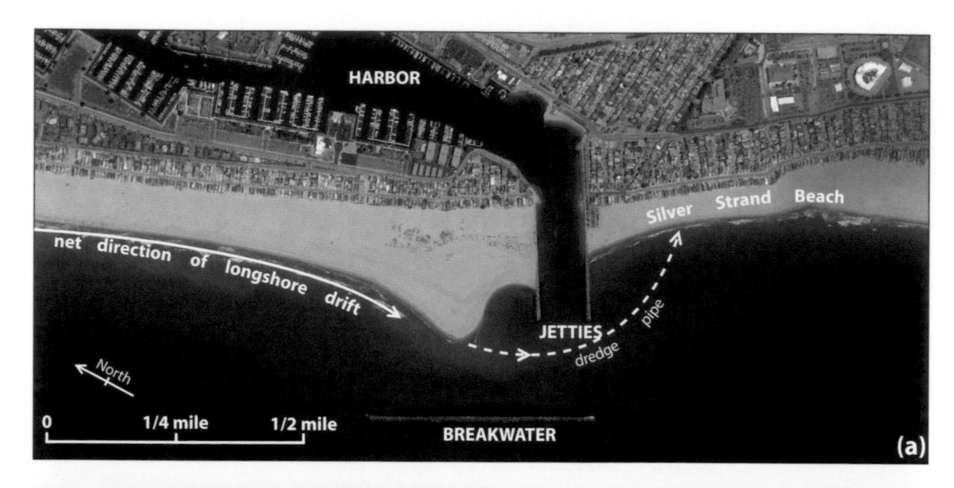
HARBOR
Silver Strand Beach
net direction of longshore drift
JETTIES
dredge pipe
North
0
1/4 mile
1/2 mile
BREAKWATER
(a)

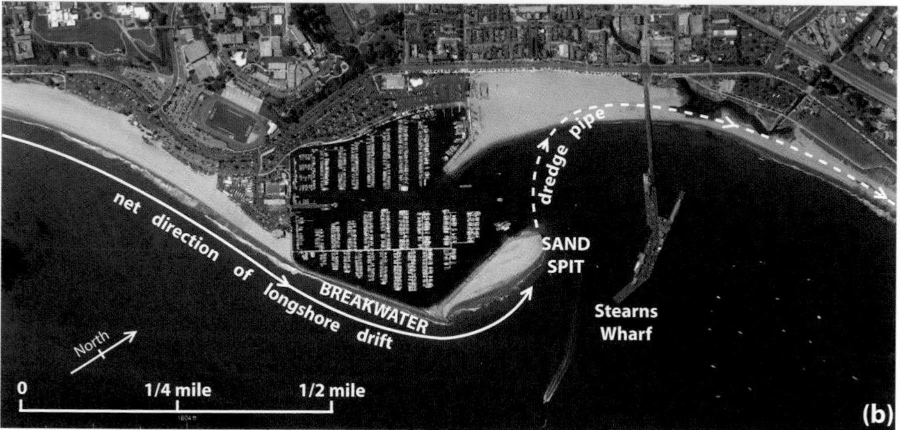
dredge pipe
net direction of longshore drift
BREAKWATER
SAND SPIT
Stearns Wharf
North
0
1/4 mile
1/2 mile
(b)

(c)

(examples: 300,000 cubic yards per year past Santa Barbara Harbor; 1 million cubic yards per year past Channel Islands Harbor near Oxnard; 50,000 cubic yards per year along the Malibu coast; and 275,000 cubic yards per year along the San Diego coast).

These reduced deposits and interrupted transfers within the beach's sand "bank account" perhaps wouldn't matter so much if sand *withdrawals* didn't continue ceaselessly. The eventual fate of most Southern California beach sand is to slide down a submarine canyon. Dozens of submarine canyons crease California's continental shelf (see text box and figures 5.8 [La Jolla Canyon] and 5.11 [Redondo Canyon]). Five of these canyons are large enough and come close enough to shore to act effectively as bottomless drains for beach sand. The sand travels south along the beaches via longshore drift until it reaches the head of a canyon, whereupon it avalanches down the canyon into deep water. The submarine canyons define the boundaries of what we call *beach compartments* (figure 6.9), each of which is like a separate bank account for sand. Deposits come in via rivers and bluff erosion. Longshore drift transfers the sand net southward. Withdrawals take place as the sand funnels down the submarine canyons. Human development has drastically cut down on sand deposits, but withdrawals continue down submarine canyons. With deposits no longer keeping pace with withdrawals, the beaches are shrinking.

Could the canyons ever fill, stopping the withdrawals and eventually stabilizing the beaches? No, the canyons are far too big. La Jolla Canyon, for example, has a volume of at least 10 billion cubic yards, so at 275,000 cubic yards per year it would take nearly forty thousand years to fill with sand. What about blocking the canyons near their heads with underwater dams to keep the sand from draining away? Legal and

---

FIGURE 6.8 (OPPOSITE). Wherever breakwaters and jetties protect harbor entrances, waves lose their power to move sand along the coast by longshore drift. If not dredged, this drifting sand soon fills harbor entrances. (a) Channel Islands Harbor near Oxnard, constructed in 1960. The large bulge in the beach shows where south-drifting sand has piled up in the calm water behind the breakwater. About one million cubic yards of sand need to be dredged each year here to keep the harbor open. (© 2014 Google; data CSUMB SFML, CA OPC; image ©2014 TerraMetrics; labels added by the author.) (b) Santa Barbara Harbor, constructed in 1930. Longshore drift moves sand past the breakwater and dumps it in the calm water of the harbor, forming a growing sand spit. The sand would plug the harbor entrance if it were not dredged at the rate of three hundred thousand cubic yards per year. (© 2014 Google; data CSUMB SFML, CA OPC; image ©2014 TerraMetrics; labels added by the author.) (c) Part of the Santa Barbara Harbor dredge pipe before it was buried below the beach. (Photograph by the author.)

California's Submarine Canyons

As I explain in this chapter, the eventual fate of most Southern California beach sand is to slide down a submarine canyon. Hundreds of submarine canyons slice across the continental shelves of the world. One of the largest is Monterey Canyon, which begins just offshore of Moss Landing in Monterey Bay (see figure opposite). A flyover of Monterey Canyon with the Pacific Ocean removed would show you a fifty-mile-long valley wider and deeper than Arizona's Grand Canyon. Several dozen submarine canyons notch the continental shelf of the Southern California Bight. Five of them are large enough and reach close enough to shore to act as major sinks for beach sand (figure 6.9).

The origin of submarine canyons puzzled oceanographers for decades after their discovery. Many canyons are sinuous and V-shaped in profile like eroded river valleys, so a sensible explanation seemed to be that the canyons were cut by rivers that flowed across the continental shelves during the low sea levels of past ice ages. Further study of the canyons and of ice age history revealed two problems with this idea. First, sea level never dropped more than about four hundred feet below its present level during the ice ages of the past several million years, yet many submarine canyons extend down thousands of feet. Second, although many submarine canyons head just offshore of where rivers end, many do not. Conclusion: Rivers during the ice ages may have cut the heads of some submarine canyons, but that leaves the rest unexplained. In 1929, a great earthquake off the coast of Nova Scotia in the North Atlantic offered tantalizing evidence for a new way to cut submarine canyons. For hours after the earthquake, one trans-Atlantic telegraph cable after another snapped across the deep seabed south of the continental shelf. The breaks occurred over a thirteen-hour span from north to south into progressively deeper water and extended more than two hundred miles. The reason turned out to be a massive underwater landslide: a dense, flowing avalanche of water, sand, and mud that cruised across the ocean floor for hours with enough force to snap undersea cables as thick as your arm.

economic issues make that a tough sell. Underwater dams would destroy acres of marine habitat, and some submarine canyons lie in legally protected habitat areas. Moreover, the trapped sand would have to be dredged back to the beach, and the cost of dredging skyrockets with water depth. Add the cost of underwater dam construction, and the expense would far exceed that of dredging sand from shallow water

Monterey Canyon. The canyon begins a few hundred feet from shore near Moss Landing in the center of Monterey Bay, and drops to a mile deep within ten miles of shore. Turbidity currents are thought to be the main force that cuts the world's submarine canyons. (Image from GeoMapApp, www.geomapapp.org.)

Oceanographers call these muddy undersea avalanches *turbidity currents*. We think that hundreds of them cascade down unstable submarine slopes every year. Turbidity currents are powerful forces of erosion. A submarine canyon might begin where a turbidity current etches a gully into an undersea slope. The process soon becomes self-reinforcing, much like a gully on land catching and funneling rainwater to enlarge itself with each storm. Today, we think that turbidity currents are the main erosional force that cuts and deepens most of the world's submarine canyons.

near the beach. (Nearshore dredging is the most common approach to beach replenishment.) For practical purposes, once a sand grain rolls down a submarine canyon, it is lost from the beach forever.

It was a glowing spring morning when I walked an abandoned dirt road to Matilija Dam. Wildflowers cloaked the valley walls, and floral scents and buzzing insects filled the air. The road contoured along the

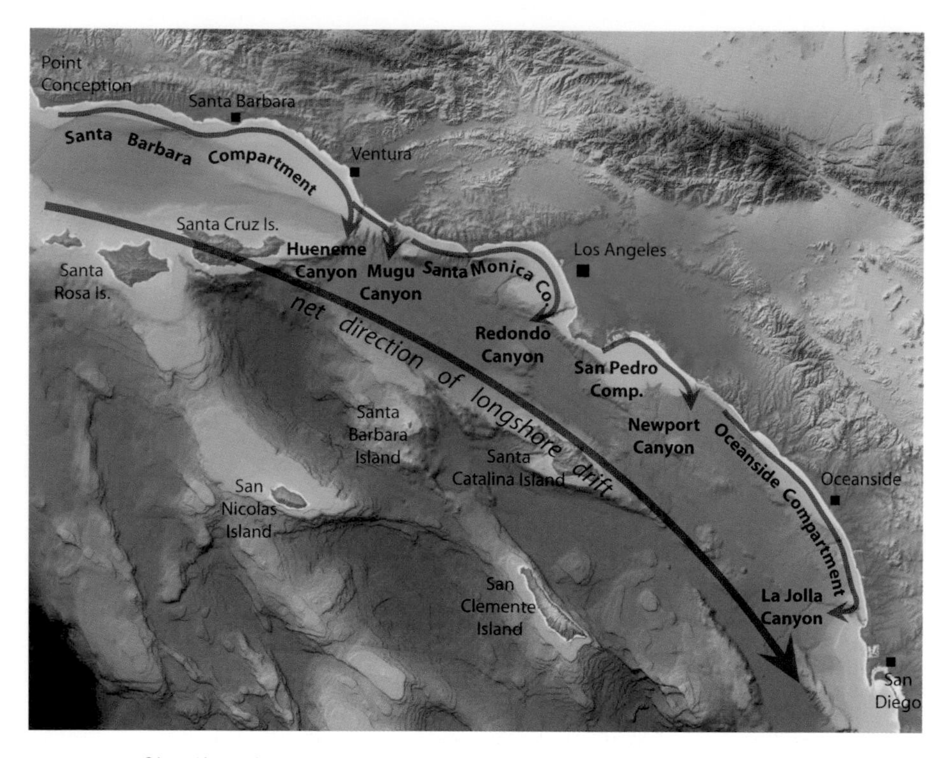

FIGURE 6.9. Gigantic undersea canyons divide Southern California's shoreline into four major beach compartments. Rivers and natural bluff erosion deliver sand to each compartment. The sand moves net southward within each compartment by longshore drift. The sand exits each compartment where it funnels down one of the canyons. (Shaded relief image from NOAA, with labels added.)

slope above the reservoir, now mostly a marsh because of siltation. An angler was trying his luck near the dam, in the 3 percent of the reservoir that still holds water instead of silt. "Will you eat it?" I asked of the runt fish he had just caught. "No, too small—fish don't grow big here anymore." He threw it back. "I think I've caught the same one three times this morning." Matilija Reservoir is now too small to grow fish either large or smart, apparently.

The dam lay in terrible disrepair, its concrete cracked and crumbling. Several streams of clear water poured over the top. After construction, fears about inferior concrete prompted authorities to cut thirty feet off the dam to reduce stress, nearly halving the reservoir's volume and, thus, its useful life span. The scene at Matilija Dam presages the fate that awaits most of the nation's dams. All dams have a finite life span set by simple math: Reservoir capacity divided by yearly sediment input equals life span in years. Obviously, dams are a key component of Southern California's water supply, and it would be hypocritical of me to bemoan them while I take showers and do laundry like everyone else who benefits from water storage. But dozens of Southern California dams are at or near the end of their useful lives. The reservoirs of some, like Rindge Dam on Malibu Creek, have long since filled in (figure 6.7b).

What should we do about obsolete dams? One option is to do nothing. Once sand has filled their reservoirs, the rivers will carry new sand over the tops of the dams and onward to the beaches. In this case, a century from now (if humans were to do nothing), a flyover of Southern California during a wet winter would show you dozens of brown, sand-filled waterfalls cascading over dozens of deteriorating dams, with the turbid rivers once again moving sand downstream to the beaches. But this isn't likely to happen, because people are going to keep living here and will probably need more water than ever. In that case, a flyover would likely show you a landscape peppered with dozens of *new* dams upstream of old ones, while downstream you would see the desiccated mudflats of silted-in reservoirs behind a bunch of deteriorating old dams, with dead riverbeds cut off, as ever, from the sea.

What about tearing down obsolete dams and letting the freed rivers carry the trapped sand downstream to the starving beaches? Matilija Dam reveals some of the complications with this idea. Ventura County resolved to remove the dam in 1998 and contracted a feasibility study with the federal Bureau of Reclamation—the nation's authority on dam construction. In the bureau's report, one problem rises above all others: how to deal with the backlog of accumulated sediment. The simplest and least expensive option—complete removal of the dam at once—would, the bureau concluded, "have a high risk potential for extreme social and environmental damage." Translation: The river, suddenly freed, and with six million cubic yards of muck at its disposal, would rampage downstream in turbid slurries, wrecking private property and entombing large swaths of riparian habitat in slurping goop. Instead, the bureau recommended that the sediment be scooped out of the reservoir. But what to do with it? Trucking it all the way to the coast would be monumentally expensive and would create traffic jams for years on the tiny mountain roads nearby. One option, the bureau suggested, would be to move the sediment downstream by slurry pipe to a storage site where it could be processed, with the large rocks and small silt separated from the beach-quality sand and sold for construction (thus recouping some of the costs). NIMBY (not in my back yard) issues have so far nixed this option; apparently no community downstream of Matilija Dam wants mountains of fetid reservoir slurry heaped up next door. Another option, the Bureau of Reclamation suggested, would be to push the sediment up onto the mountain slopes upstream of the dam and stabilize it before removing the dam. This option would let the river run free, but it wouldn't deliver much of the trapped sand to the beaches, and it would plaster muck across acres of natural mountainside. Other proposals involve incrementally notching the dam a few feet at a time, allowing the river to move the sediment downstream in pulses. Depending on the river's efficiency, this could take decades. Today, despite multiple studies and proposals, no solution has yet satisfied enough constituents, and Matilija Dam still stands. The only visible sign of progress is a "cut here" graffiti stripe painted on the dam's two-hundred-foot-high face (figure 6.7a).

While some dams may be successfully removed in coming decades, I suspect that widespread dam removal won't become a feasible part of beach restoration in Southern California any time soon. Dams are too integral to our water supply, and many dams still have decades of useful life. And as Matilija Dam shows, removing obsolete dams and dealing with the backlog of sediment is not a simple task. Since many Southern California beaches get a lot of their sand from bluff erosion, tearing down seawalls and allowing bluffs to erode naturally (as part of a strategy of managed retreat) would help some beaches. But exactly the opposite is happening. New seawalls go up every year, and few old seawalls are ever removed. I think Southern Californians have only one viable option when it comes to rebuilding sand-starved beaches: committing millions of dollars to repeated and sustained *beach replenishment*—importing sand to depleted beaches.

During the economic boom years after World War II, a number of major coastal construction projects contributed huge amounts of sand to local Southern California beaches. Between the 1940s and the 1960s, several large marinas and harbors were dredged out of the coastline—including Mission Bay, Oceanside Harbor, Marina Del Rey, and Channel Islands Harbor—and the dredged sand was placed on nearby beaches. The beaches of Los Angeles also gained sand from construction of the Los Angeles International Airport, the El Segundo Oil Refinery, and the Hyperion Sewage Treatment Plant. The increases in beach size from these and other construction projects lasted for several decades. But by the 1970s, most major sand-generating construction projects had been completed. Sand replenishments from these sources tapered off, while seawalls continued to go up and dams continued to impound river sand. Large storm waves and higher sea levels during several El Niño–Southern Oscillation episodes* accelerated sand losses, particularly during 1982–83, 1988–89, and 1997–98. By the late 1990s, sand losses were clearly evident at shrunken, rocky beaches throughout much of Southern California.

I moved to the San Diego area in 1997, and I remember taking my first field trips with students to the El Niño storm-ravaged local beaches. Narrow strips of sand faded into long stretches of cobbles that often allowed no dry passage along the bluffs, even at low tide. In 2001, governments in San Diego's coastal communities—prompted by fears of reduced tourist revenue from shrunken beaches and property damage from accelerating bluff erosion—banded together to launch the San Diego Regional Beach Sand Project. Funded with $17.5 million in federal, state, and local government revenue, the project pumped 2.1 million cubic yards of sand from several offshore dredge sites onto the fifty-mile stretch of beaches between Oceanside and the Mexican border. (For perspective, 2.1 million cubic yards of sand would fill one average-sized football stadium slightly beyond the brim.)

Follow-up measurements revealed great beach-to-beach variability in how long the infusion of dredged sand lasted. Some beaches lost most of their sand within two years; others retained significant sand for more than six years. Within a decade, though, practically all the beaches had shrunk to prereplenishment levels. This should come as little surprise. La Jolla Canyon (figure 6.9) funnels 275,000 cubic yards of south-drifting sand away from San Diego's beaches every year (the amount delivered to the canyon by southward longshore drift). Roughly speaking, the 2.1 million cubic yards of sand added in 2001 had an expected life span of 2.1 million cubic yards ÷ 275,000 cubic yards per year lost down canyon = 7.6 years. And therein lies the reality of beach replenishment. As long as river dams and seawalls curtail deposits to the beach's sand bank account, and as long as canyons continue to siphon away sand, artificial replenishment has to happen repeatedly to sustain the beaches.

In 2012, local San Diego governments funded another beach replenishment project, similar to the 2001 project although it pumped less sand (1.5 million cubic yards). A key difference this time was that surveyors

* El Niño–Southern Oscillation conditions (commonly known as ENSO) occur when periodic changes in atmospheric circulation allow warm waters from the western tropical Pacific near Indonesia to flow east across the ocean to the coasts of North America and South America. The warm waters raise coastal sea levels by nearly a foot and provide energy for more storms, both of which increase beach erosion.

identified dredge sites that contained coarser-grained sand (sand made of larger pieces). The hope is that the larger particles will persist longer on the beaches.

Is rebuilding sand-starved beaches worth the money? It depends on what assumptions you make about value added from tourism, tax revenue, property protection, and quality of life. My investigations of the issue have convinced me that, in San Diego at least, revenue enhancements from bigger beaches justify replenishment costs. Others may reach a different conclusion, but one thing seems certain: As long as river dams, seawalls, breakwaters, and jetties upset the natural balance of sand, Southern Californians will have to either live with shrinking beaches or regularly spend millions of dollars to do the work that Nature did before.

7

Sea-Level Changes and the Ice Ages

It is mildly unnerving to reflect that the whole of meaningful human history—the development of farming, the creation of towns, the rise of mathematics and writing and science and all the rest—has taken place within an atypical patch of fair weather.

—Bill Bryson, *A Short History of Nearly Everything*

Half a mile from my house and three miles from the beach in northern San Diego County is a mesa, three hundred feet above the sea, cloaked with chaparral. The plants send their roots into ground packed with smooth, round boulders (figure 7.1). There are two main ways that nature makes boulders smooth and round: by tumbling them along riverbeds or rolling them on beaches. Here it was a beach—but one where waves last broke seven hundred thousand years ago. When saltwater last splashed these rocks, our *Homo erectus* ancestors in Africa were still figuring out how to use fire.

The mesa is one in a series that rises inland across San Diego like a staircase from the coast to the mountains. Beach sands, wave-rolled boulders, and occasional fossil seashells pave each mesa, telling us that waves cut each one at a former shoreline. We can see the same thing at the shore today. As I pointed out in chapter 6, sea level has barely budged in the past seven thousand years, which means that waves worldwide have been hurling themselves against the continents at about the same level this whole time. As a result, wherever bedrock and shore-break meet, waves have cut a large step into the land, consisting of a wave-cut platform (the flat part of the step) capped by a thin beach and backed by a sea cliff (figure 7.2). More than half of Southern California's coast looks this way.

FIGURE 7.1. My good dog Storm on a marine terrace—an old shoreline high above the sea in northern San Diego County. Waves rolled these boulders into rounded shapes about seven hundred thousand years ago. Since then, earthquakes have incrementally hoisted the land so that the terrace now lies three hundred feet above sea level. The terrace spreads as a mesa, acres across, and is one of more than a dozen terraces that ascend inland like a staircase from the coast to the mountains. (Photograph by the author.)

The staircases of old wave-cut platforms and sea cliffs that ascend inland from the coast form natural construction sites and are crowded today with homes and businesses. In San Diego, the communities of La Mesa, Mira Mesa, Kearny Mesa, Clairemont Mesa, Otay Mesa, and others take their names from the wave-cut staircase. Likewise, Costa Mesa in Orange County, Mesa Street in San Pedro, Mesa Park in Santa Barbara, and numberless other "mesa" names throughout coastal Southern California reflect this staircase geography. Development has obscured the staircase along parts of the mainland, but if you head out to the islands, you'll find pristine examples, especially on the southwestern face of San Clemente Island (figure 7.3). Crowd that mesa staircase on San Clemente with Starbucks, McDonald's, supermarkets, and suburban houses, and you'll have duplicated much of Southern California's coast.

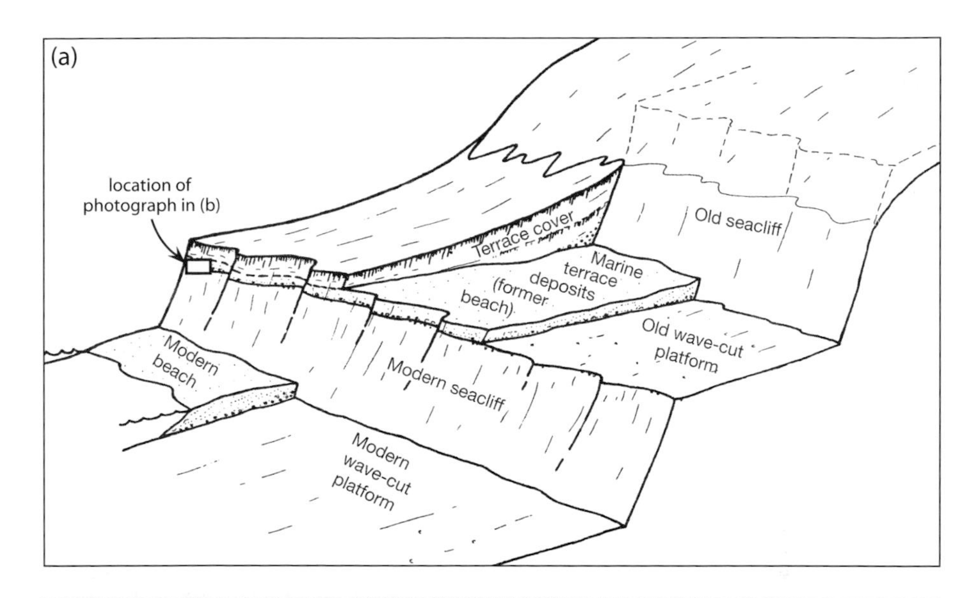

FIGURE 7.2.
(a) Marine terraces are old wave-cut platforms, now above the sea, and exhibit all the features we see on modern wave-cut platforms. These include a nearly flat erosion surface that waves have cut onto bedrock, paved with a veneer of wave-rounded pebbles, seashells, and beach sands, and backed by a sea cliff. Once a wave-cut platform emerges from the sea, rain and streams wash debris down onto it, burying the marine deposits while wearing down the former sea cliff. (Adapted from Harden 2004: fig. 15–19.)
(b) Wave-polished boulders and fossil clamshells on top of a marine terrace in Torrey Pines State Park near San Diego (note their location in panel a). By age-dating the fossils in terrace deposits, we can work out the timing of terrace formation and uplift. (Photograph by the author.)

FIGURE 7.3. The southwest face of San Clemente Island displays some of the most clear-cut marine terraces in Southern California. Crowd these terraces with houses and businesses, and you have a fair image of what most of the mainland coast looks like. (Photograph by Dan Muhs, U.S. Geological Survey.)

Geologists call these old wave-cut mesas *marine terraces*. Wherever you see a staircase of marine terraces, you'll nearly always find that it has a predictable age-progression. The lowest terrace is usually the youngest, and the terraces get progressively older as you ascend the staircase. Figure 7.4 illustrates this for San Diego's marine terraces. Notice that the lowest terrace is the youngest at about forty-five thousand years, and the highest is the oldest at more than 1.5 million years. This age-progression tells us something about how marine terraces form. But before I get into that story, let me explain how we can figure out the ages of marine terraces.

Geologists take advantage of several natural clocks to determine the ages of rocks and fossils. One set of clocks is based on radioactive decay—the natural and predictable change of unstable atoms into stable forms. For instance, uranium-238 (^{238}U)—one of several isotopes of

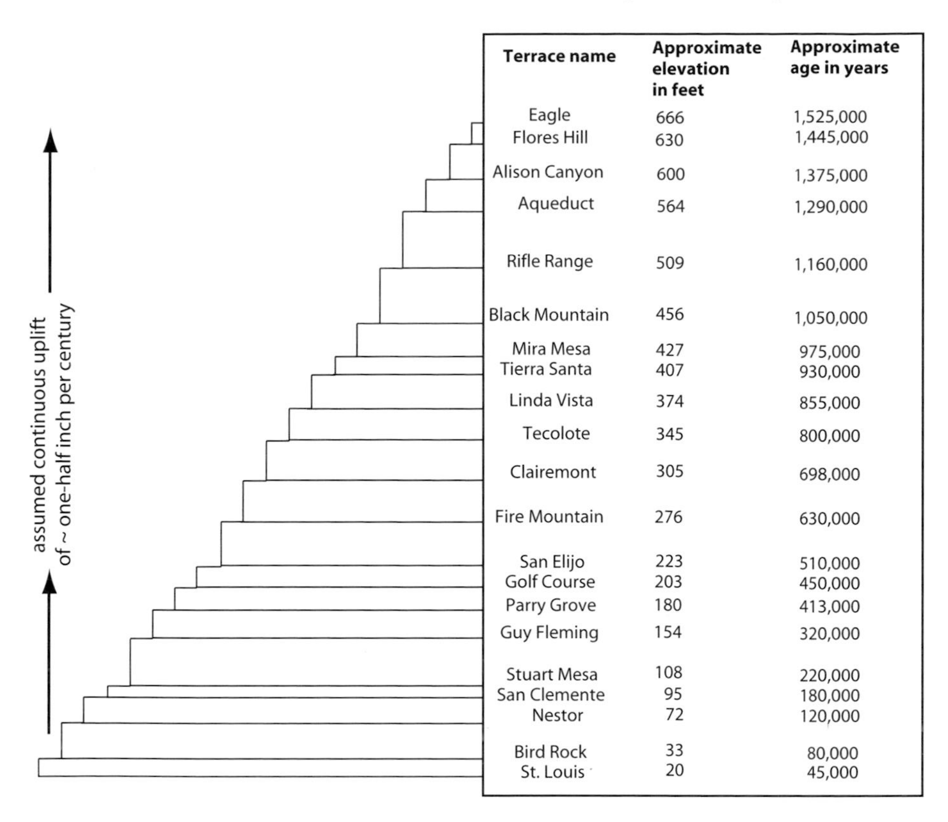

Terrace name	Approximate elevation in feet	Approximate age in years
Eagle	666	1,525,000
Flores Hill	630	1,445,000
Alison Canyon	600	1,375,000
Aqueduct	564	1,290,000
Rifle Range	509	1,160,000
Black Mountain	456	1,050,000
Mira Mesa	427	975,000
Tierra Santa	407	930,000
Linda Vista	374	855,000
Tecolote	345	800,000
Clairemont	305	698,000
Fire Mountain	276	630,000
San Elijo	223	510,000
Golf Course	203	450,000
Parry Grove	180	413,000
Guy Fleming	154	320,000
Stuart Mesa	108	220,000
San Clemente	95	180,000
Nestor	72	120,000
Bird Rock	33	80,000
St. Louis	20	45,000

FIGURE 7.4. The age and average elevation of marine terraces in San Diego County. Notice that the terraces become older as they ascend—a feature common to all of Southern California's marine terraces. Most terraces formed during sea-level highstands at the peak of past interglacial periods. Since we know the level to which the sea rose during many of these past highstands (see figure 7.5a), we can use the age and current elevation of the terraces to calculate rates of uplift (i.e., how fast the land is rising). The terraces of San Diego indicate an average uplift rate of about a half-inch per century, but other areas of the Southern California coast are rising much faster—as much as fifty inches per century near Ventura, for example, because of more frequent fault movements there. The diagram vastly exaggerates the steepness of the land. Individual terraces are often more than a mile wide but usually just a few tens of feet apart in elevation. The elevations listed are averages; the height of any terrace may vary widely across its expanse as a result of fault movements and tilting of the land. (Based on Kennedy and Tan 2005a and 2005b.)

uranium—changes steadily over time into several other types of atoms. (Isotopes are varieties of an element created by different numbers of neutrons in the atomic nucleus.) When radioactive atoms change, or "decay," little particles jet out from the atomic nucleus at high speeds. (You don't want to be near concentrated amounts of some types of radioactive atoms because when these high-speed subatomic particles shoot out, they tear through your flesh like tiny bullets, burning tissues and sometimes turning cells cancerous. Happily, the natural radioactivity in most rocks and fossils is far too low to pose a health threat.) The key thing about radioactivity for geologic dating is that radioactive atoms change in a predictable, clocklike way.

One radioactive element that has proved spectacularly useful for dating organic remains—wood, teeth, bones, and shells in particular—is carbon-14 (^{14}C). Carbon atoms occur in three varieties, or isotopes: ^{12}C, ^{13}C, and ^{14}C. The first two are stable, but ^{14}C is radioactive. It forms in the upper atmosphere as the Sun's rays bombard atoms of Nitrogen-14 (^{14}N), transforming some of them into ^{14}C. The ^{14}C then changes back into ^{14}N at a precise rate. As plants grow, they add all three types of carbon to their tissues. Likewise, as animals grow, they eat and drink all three forms of carbon. Every living creature thus has some ^{14}C in its tissues. Once a creature dies, though, it stops adding carbon to its body. The ^{14}C already there transforms steadily into ^{14}N. The longer a creature has been dead, the less ^{14}C there is in its tissues. The relationship between the amount of ^{14}C remaining and the time since death is quite precise, allowing us to calculate radiocarbon ages with high accuracy. Wood, teeth, bones, and shells can endure for thousands of years and are thus the materials we most commonly date using ^{14}C.

A drawback of ^{14}C is that it works only to ages of about seventy thousand years. In organic remains older than that, any ^{14}C left over is of such low concentration as to be unmeasurable. Seventy thousand years is fine for some types of work, including most archeological studies. But most of Southern California's marine terraces fall beyond the range of ^{14}C. To date older terraces, we turn to two other tools: uranium–thorium decay and amino acid racemization.

The element uranium occurs in minute concentrations in seawater, and certain types of organisms, particularly corals, absorb it as they grow. One isotope of uranium, ^{238}U, decays through a series of steps to ^{230}Th, an isotope of thorium. As a coral grows, it adds tiny amounts of ^{238}U to its skeleton. Over time, this ^{238}U steadily transforms into ^{230}Th. The proportion of the two isotopes changes in a predictable way over time, allowing us to calculate the ages of fossil corals in marine terraces back as far as five hundred thousand years.

A drawback of uranium–thorium dating is that it doesn't work on most fossils. Shells of molluscs like clams and snails are common in marine terrace deposits, but molluscs don't take up uranium from seawater. Here, though, we have another trick: amino acid racemization.

The proteins of living creatures contain amino acids in a specific molecular shape known as the *L-configuration.* Upon death, some of these amino acids begin shape-shifting to a new arrangement called the *D-configuration*—a process called *racemization.* Molluscs are particularly useful for amino acid dating because they live practically everywhere in shallow ocean habitats and because their shells are bound together with matrixes of amino acid–rich proteins. By measuring the ratio of the two types of amino acids in mollusc shell proteins, we can estimate the time since death. One drawback is that the rate of racemization depends on temperature—the shape-shift happens faster at higher temperatures—so we have to make realistic assumptions about the temperatures experienced by shells in terrace deposits.

Coming back now to Southern California's marine terraces, let's figure out why, in a staircase of terraces, each step gets progressively older as you ascend. (Figure 7.4 shows this for San Diego's marine terraces.) You might reasonably imagine that marine terraces formed at times when the sea was higher than it is today and were cut, one after another, as the sea incrementally dropped, sort of like a series of soap rings left on the walls of a draining bathtub. But two facts will convince you that this can't be right. First, sea level during the past few million years has mostly been *lower* than it is today (see figure 7.5a) as the Earth has cycled through multiple ice ages—a topic to which I'll return in a moment. Second, if all the glacial ice on Earth today were to melt, sea level would rise about 230 feet, and many marine terraces lie much higher than this. (The highest terrace on San Clemente Island, for example, stands nearly a thousand feet above the sea.) Moreover, terraces of the same age often occur at different elevations in different places, especially on the opposite sides of active faults. This tells us that movements of the land, not just the sea, must be a key part of the terrace-formation story. Southern California, sandwiched as it is between moving tectonic plates, is broken up into dozens of crustal blocks that shift up, down, or sideways with every earthquake. Any marine terraces that we see above sea level today invariably occur on *rising* blocks of crust. These include San Clemente Island, San Nicolas Island, the San Diego coast, the Palos Verdes Hills, the Ventura and Santa Barbara coasts, and the northern Channel Islands. Figure 7.6 shows an example of uplifted marine terraces in the Palos Verdes Hills.

The full story, illustrated in figure 7.7, involves the up-and-down yo-yoing of the sea superimposed onto rising blocks of land. To form a terrace, ocean waves need to hack away at the same place against the

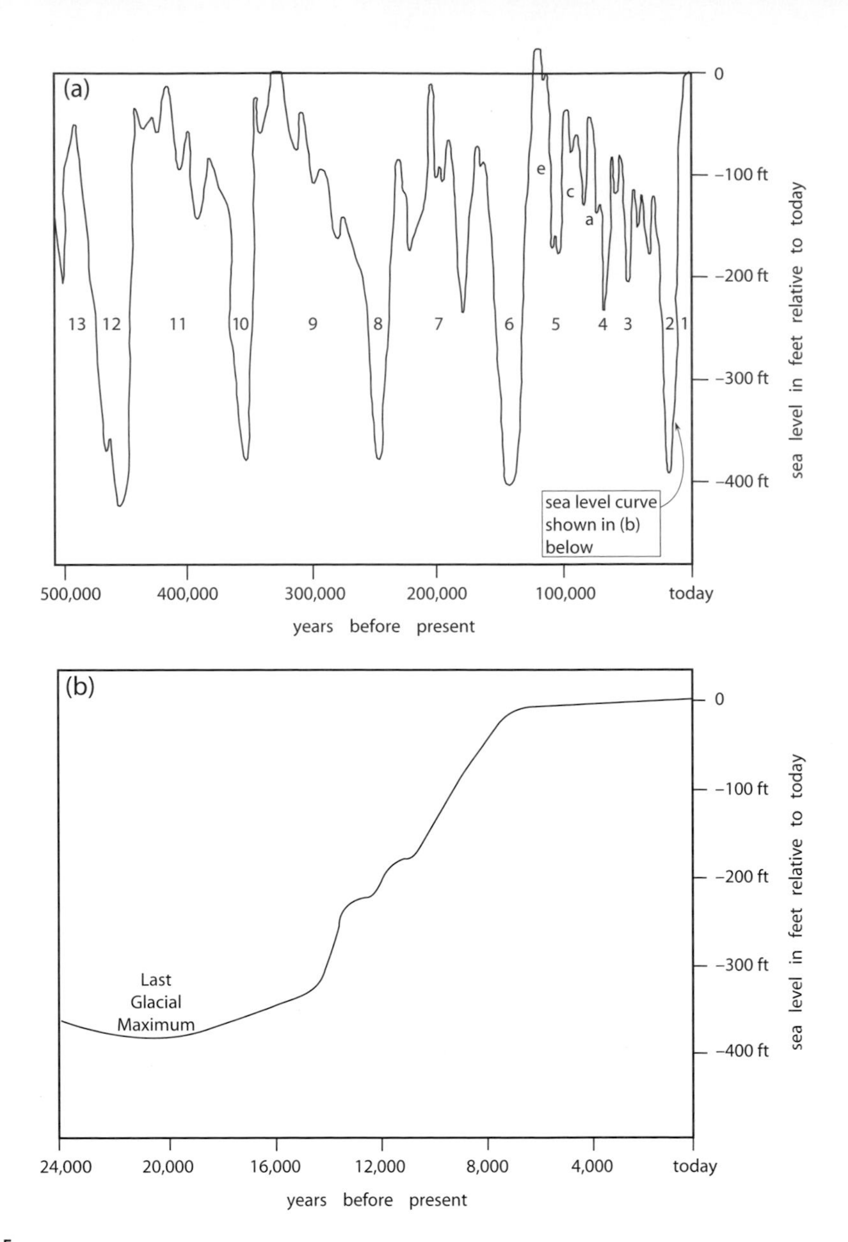

FIGURE 7.5
(a) Sea-level curve for the past five hundred thousand years, based on oxygen isotope data from fossils in deep-sea sediment cores. You can see that sea level has risen and fallen on a repeating cycle of about a hundred thousand years, with shorter cycles imposed on that long cycle. During the highest highstands, sea level typically rose to within a few feet of where it is today. During the lowest lowstands, the sea dropped close to four hundred feet below where it is today. The numbers and letters indicate oxygen isotope stages. Notice that during stage 5e, about 120,000 years ago, sea level was roughly twenty feet higher than it is today. The marine terrace that formed during this highstand is very prominent along much of Southern California's coast; Highway 101 and many coastal communities are built on it. (From Kennedy and Tan 2005a and 2005b.) (b) Sea-level curve for the past twenty-four thousand years, illustrating the sea's rise from the lowstand of the most recent glacial period—known as the Last Glacial Maximum—to today's interglacial highstand. Notice that the sea had risen to nearly its present level by about seven thousand years ago. The stability of this interval—a span that encompasses most of recorded human history—is clearly not typical of the past. (Based on Peltier and Fairbanks 2006.)

FIGURE 7.6. Marine terraces of the Palos Verdes Hills. The view is east across Point Vicente, with the Point Vicente lighthouse in the foreground. The lowest and most prominent terrace (marked by the two rightmost arrows) was cut during an interglacial highstand about eighty thousand years ago, when sea level was forty to sixty feet lower than it is today. (This highstand corresponds to oxygen isotope substage 5a in figure 7.5a.) Uplift of the land has since raised the terrace to its current elevation about 140 feet above the sea, which translates to an average uplift rate of close to three inches per century—a geologically rapid pace due to the activity of the Palos Verdes fault and other nearby faults, all of which together have raised the Palos Verdes Hills. Other arrows mark higher terraces cut during older highstands. For an explanation of how staircases of marine terraces form, see figure 7.7. (Photograph © 2002–2014 Kenneth and Gabrielle Adelman, California Coastal Records Project, www.californiacoastline.org.)

land for at least a few millennia. Most marine terraces therefore form during what we call *highstands*—long periods of high sea level when the Earth's polar ice caps had mostly melted—or *lowstands*—long periods of low sea level at the peak of glacial intervals when polar ice caps had grown to maximum size. We are currently in the highstand of the present interglacial period (figure 7.5b), which means that terraces cut during prior lowstands are now mostly underwater. The terraces we see

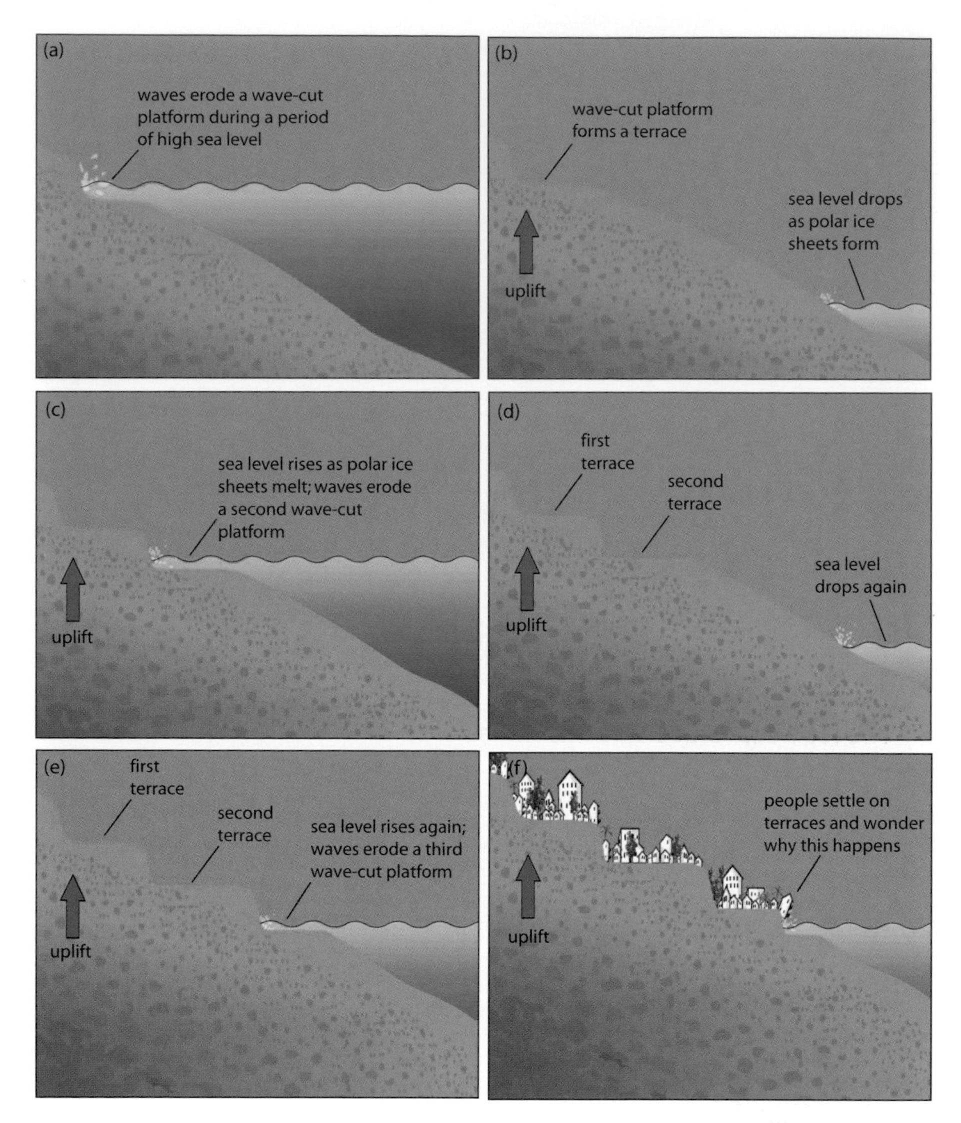

FIGURE 7.7. Marine terraces form when the up-and-down cycles of the sea (caused by the waxing and waning of polar ice sheets) are superimposed onto rising blocks of land. (a) Most terraces are cut during *highstands*—periods of high sea level when polar ice sheets have shrunk to their minimum size. (b) When polar ice sheets reform, sea level drops; meanwhile, the land rises incrementally as the crust shifts during earthquakes. (c) When the sea returns during the next highstand, the earlier terrace has risen beyond reach, so the sea cuts a second terrace below the first one. (d) The sea drops again, and the land continues to rise. (e) When the sea returns to yet another highstand, it cuts a third terrace below the first two. (f) Many Southern California coastal communities are built on marine terraces. (Adapted from an animation by Tanya Atwater, University of California, Santa Barbara.)

above the sea today were cut during previous interglacial highstands. The sequence—and here you might want to follow along in figure 7.7—goes like this: Waves cut a terrace during a highstand (7.7a). The sea drops as polar ice caps regrow, and meanwhile the land rises (7.7b). The sea returns to another highstand, but the first terrace has now risen beyond its reach, so the sea cuts a second terrace below the first one (7.7c). The sea drops again while the land continues to rise (7.7d). When the sea rises again to yet another highstand, uplift has raised the earlier terraces beyond its reach, so the sea cuts the next terrace below the previous ones (7.7e). Repeating the cycle produces a staircase of marine terraces that get progressively older going upward (7.7f; see also figure 7.4).

As so often happens in science, answering one question—how marine terraces form—brings up several others. How do we know about the ice ages? How do we figure out the timing of the ice ages, and the dramatic swings in sea level they produce? And what causes the ice ages? So here I want to take you on a rather long aside to explore one of the great stories of modern science: the discovery of the ice ages and their cause.

THE ICE AGES: DISCOVERY AND CAUSE

For about the past seven thousand years, humanity, with few exceptions, has enjoyed relatively stable climates and sea level. Most of the human history that any of us ever learned took place during this span, so anyone could be forgiven for assuming that such stability is normal. But if we expand our view just a little, to encompass the past two to three million years, it's clear that the stability of recent millennia is an anomaly. Imagine the stock market holding steady for a couple of weeks, with prices neither rising nor falling. Would you conclude that such stability is normal? Of course not, because a longer view shows that stock prices can fluctuate wildly. Likewise, the climate stability of recent millennia doesn't reflect the natural volatility of the Earth's climate over longer spans.

During the past two to three million years, the Earth has swung, pendulum-like, between two climate extremes: cold glacial periods, or ice ages, when polar ice sheets have advanced across the continents like pancake batter poured into a pan, and warmer interglacial periods when the ice sheets have melted away. There have been roughly thirty glacial–interglacial cycles during this span. The glacial intervals have lasted longer—about fifty thousand years on average—whereas the interglacial periods have averaged about ten thousand years. In some sense, the warm interglacial periods—including the present one—are aberrations imposed onto the dominantly glacial climates that have persisted on Earth for most of the past two million years. At the peaks of the glacial periods, ice sheets more than two miles thick covered much of Canada, northern Europe, and Siberia. The ice sheets grew by gathering snow that started as water evaporated from the oceans. As the ice expanded, the sea dropped (by as much as four hundred feet during the coldest glacial cycles, figure 7.5a). The most recent swing of the climate pendulum brought the Earth from the peak of the last glacial period about twenty thousand years ago—a time known as the Last Glacial Maximum—into the present interglacial period (figure 7.5b). The ice, even at its maximum extent, never came to Southern California. But if you visit Scandinavia, Canada, the Rockies, the Cascades, or the high Sierra Nevada, you'll see the signatures of departed ice all around you: immense piles of rock rubble, called *moraines,* bulldozed by the ice; mighty scratches left where the ice dragged boulders across the bedrock; and immense U-shaped valleys—including Yosemite Valley and Kings Canyon—with their walls scraped and polished by passing ice.

The signature of past ice ages may seem clear today, but their recognition and, ultimately, the discovery of their cause spanned several scientific generations. The famed explorer Alexander von Humboldt once observed that there are three stages to scientific discovery: first, people deny it is true; then they deny it is important; finally, they credit the wrong person. Humboldt may have had in mind the career of a nineteenth-century Swiss naturalist named Louis Agassiz.

For nineteenth-century scientists, much of northern Europe's landscape, particularly Switzerland's, presented a singular puzzle. Immense boulders, some as large as houses, lay haphazardly about the fields and valleys. They were made of rock unlike the local bedrock, but much like the rock in distant mountains. They were dubbed *erratic boulders,* or "erratics." One naturalist, Jean André de Luc, proposed that pressurized air in distant caves had launched the erratics like artillery from the mountains. His flatulence theory went nowhere, but at least it was worth a laugh. A more sober proposal held that melting icebergs had dropped the erratics at a time when oceans covered the Earth. This theory faded as geologists found no evidence for a globe-covering ocean. The idea that glaciers in the Alps had once been much larger, and had carried the boulders down from the mountains, first took hold among people who lived near glaciers. These included the naturalists Karl

Schimper, Jean-Pierre Perraudin, Ignaz Venetz, and particularly Jean de Charpentier, the director of a salt mine in the tiny Swiss town of Bex. De Charpentier saw that not only erratics, but also long, arc-shaped ridges of gravel (moraines) could be explained if Alpine glaciers had once advanced far downhill, pushing the gravel like plows. He also found parallel scratches and grooves on the valley walls and floors around Bex; these too seemed to be the marks of passing ice. De Charpentier never did much to promote his theory, but he was a convivial man who frequently hosted men of learning at his home in Bex. In the summer of 1836, he hosted Louis Agassiz and his family. Agassiz (figure 7.8), although only twenty-nine at the time, had already made a scientific name for himself with award-winning publications on fossil fish. De Charpentier gave Agassiz a tour of his evidence for ice. The visit changed Agassiz's life.

True to Humboldt's third maxim—that scientific discoveries credit the wrong person—Agassiz is the person that history usually credits with the discovery of the ice ages. He neither discovered them nor coined the term *ice age* (that came from his friend Karl Schimper). But he did bring force of personality, tireless promotion, and a remarkable book. In 1840, four years after he first visited Bex, Agassiz published *Studies on Glaciers*. True to Agassiz's personal maxim—to learn by doing—the book is a masterpiece of observational science, packed with detailed descriptions and painstaking sketches from long summers spent tramping in the mountains. He matched careful descriptions of what he saw happening in modern glaciers with signs he observed in the valleys far below, assembling an edifice of evidence to back up his revolutionary claim that ice had once spread far and wide across the Northern Hemisphere. In some ways, *Studies on Glaciers* presaged a better-known masterwork of observational science published nineteen years later—Charles Darwin's *On the Origin of Species,* which likewise brought a thick stack of evidence to bear on a revolutionary idea. (Curiously, though, Agassiz never accepted Darwin's evidence for evolution.)

Despite its brilliance, *Studies on Glaciers* didn't enjoy a friendly reception at first. Agassiz stumped for his theory on both sides of the Atlantic but did himself no favors by pushing his ideas about ice ages far beyond where the evidence could reasonably take them. It was one thing to accept that mountain glaciers had once reached far down their valleys. It was another to envision what Agassiz proposed—that the Northern Hemisphere once lay fully encased in ice from the North Pole all the way to the Mediterranean. This ice, he claimed, formed a suffocating

FIGURE 7.8
Top: Swiss geologist Louis Agassiz, whose 1840 book on glaciers—along with his charisma and tireless promotion—eventually convinced the scientific community that ice ages happened.
Middle: Scottish mathematician James Croll developed the first orbital theory of ice ages.
Bottom: Serbian mathematician Milutin Milankovitch put Croll's theory on a modern footing and generally receives credit for the orbital theory of ice ages, or what we now call *Milankovitch cycles.*
(Images from Wikimedia Commons.)

blanket that had killed off practically all life across the Northern Hemisphere, including the hairy mammoths now found melting out of retreating glaciers in Siberia, their meat hacked off by local villagers to feed their dogs. It didn't help that Agassiz had no explanation for how and why such a mass of ice came and went. Science is about explanations, not just evidence, and Agassiz had only evidence.

Nonetheless, the evidence was impressive, and within two decades of *Studies on Glaciers,* most members of the scientific community had reached a middle ground between denial of ice ages, on the one hand, and Agassiz's vision of an all-encompassing freeze-over on the other. Moreover, as geologists mapped the bouldery moraines left behind by the ice, it became clear that major ice sheets had come and gone not once, but several times, across both northern Europe and North America. By the 1860s, the question was no longer *whether* the Earth had ice ages, but *why*. Proposals included the notion that the Earth passed periodically through warm and cold regions of space; that the Sun's energy waxed and waned; and that the continents shifted position, sometimes gathering near the poles to be iced over before drifting toward warmer latitudes. (This last proposal was prescient; continental drift would become an established fact in the twentieth century. But continental movements are far too slow to explain the ice ages of the past two million years. Had you looked at the Earth from space two million years ago, you would have seen the landmasses in nearly the same position as today.) The puzzle of the ice ages, it turns out, would not be solved outdoors with fieldwork, Agassiz-style, but rather indoors, with pencils, slide rules, and the laws of physics.

During the 1860s, British scientific journals began receiving papers from a James Croll (figure 7.8), who listed his address as Anderson College in Glasgow. One paper, published in the *Philosophical Magazine* in 1864, quickly drew attention as a work of impressive rigor and originality. The paper proposed that variations in the Earth's orbit caused the planet to cycle in and out of ice ages. The distance from London to Glasgow being a slog in those days, some time passed before British intelligentsia learned that Croll was not a professor at the college, but a custodian. Croll grew up the son of an impoverished Scottish crofter, and his formal education ended at age thirteen. But each day, after he had finished mopping floors at the college, Croll would hole up in the library. There, self-taught, he came to master philosophy, physics, chemistry, and mathematics. From his mind sprouted a theory of the ice ages that would be largely vindicated—albeit a century later.

Croll reasoned that the growth or shrinkage of continental ice sheets reflected a simple imbalance: If more snow fell at the poles in winter than melted in summer, the ice would advance; if more ice melted in summer than was replaced by snow the next winter, the ice would retreat. Croll thought that small shifts in the Earth's orbit might tilt things one way or the other by changing how much solar energy the Earth receives. The Earth doesn't orbit the Sun in isolation; the gravity of other planets tugs on it, warping its orbit and tweaking its axis. Venus, Mars, Saturn, and particularly massive Jupiter all tug on the Earth in cycles that Croll could calculate using laws of planetary motion. The calculations were not straightforward. Today, you would use a computer. Croll had pencil and paper. "Little did I suspect, at the time when I made this resolution," he later wrote, "that it would become a path so entangled that fully twenty years would elapse before I could get out of it." Few among us have what it takes to probe at one math problem for twenty years, but the task seems to have suited Croll's obsessive personality. Ultimately, his calculations of orbital changes extended back three million years and forward one million years.

One upshot of Croll's work was a hundred-thousand-year cycle in the shape of the Earth's orbit. The Earth's path around the Sun is not a circle but an ellipse. This means that the Earth's distance from the Sun varies throughout the year.* Croll's calculations showed that the Earth's orbit shifts from more circular to more elliptical, and back again, on a hundred-thousand-year cycle. The change is summed up by a measure called *eccentricity:* the more elliptical the orbit, the higher the eccentricity. Croll reasoned that ice ages would most likely happen at times of greatest eccentricity every one hundred thousand years.

Good scientific theories make testable predictions, and Croll's did exactly that. The problem was, science would not find a way to accurately date the ice ages for close to a century beyond Croll's day. Moreover, Croll soon ran into a bigger dilemma. His eccentricity calculations indicated that the latest ice age (what we now call the Last Glacial Maximum) should have ended about eighty thousand years ago. Yet geolo-

* Presently, the Earth's elliptical orbit puts it closest to the Sun in January (91,445,000 miles) and farthest away in July (94,555,000 miles). If this seems odd from the perspective of the Northern Hemisphere's seasons, I should point out that the winter–summer seasonal cycle is caused not by how close we are to the Sun, but rather by the tilt of Earth's axis. That tilt means the Northern Hemisphere receives less solar energy in January, even though the Earth is closer to the Sun then, and more energy in July when it's farther away.

gists were convinced that it had ended far more recently. A major piece of evidence came from Niagara Falls.

As water plummets over Niagara Falls, the eroding lip of the falls retreats upstream about three feet every year. (If you look at Niagara photographs from a hundred years ago, you'll see that the fall-line lay about three hundred feet downstream of where it is today, which translates to three feet of erosion per year.) Seven-mile-long Niagara Gorge, downstream of the falls, was clearly cut by the falls' upstream retreat, yet the Niagara River could not have begun to cut the gorge until the last ice age ended, because before that the area lay crushed under a mile of ice. But Niagara Gorge is only seven miles long, which suggests that, at three feet of erosion per year, the gorge isn't likely to be much older than twelve thousand years. It's not realistic to imagine a long delay between when the ice retreated and when the Niagara River began to flow; the melting ice would have provided massive amounts of water, quickly filling the newly formed Great Lakes, which lie in depressions scooped out by the ice. (The Niagara River takes Lake Erie's overflow to Lake Ontario.) To nineteenth-century geologists, Niagara Gorge could mean only one thing: The last ice age had ended close to twelve thousand years ago (roughly the date we accept today). Bottom line: Because of strong evidence that the last ice age had ended close to twelve thousand years ago, instead of eighty thousand years ago as Croll had predicted, Croll's orbital theory fell out of favor.

But not for long. In 1912, about two decades after Croll's death, Milutin Milankovitch, a young Serbian professor at the University of Belgrade (figure 7.8), revived Croll's orbital theory and gave it stronger legs. Milankovitch's goal was not, at first, to solve the problem of the ice ages. It was to develop a mathematical theory of climate that could predict the Earth's temperature at any location and latitude. He was immersed in that problem in 1914, when World War I broke out. Milankovitch was living in Budapest at the time, and his status as a reservist in the Serbian army got him shipped off to a prisoner-of-war camp. But Milankovitch's Hungarian colleagues soon convinced the local authorities that his scientific research was harmless, and he was released and placed under loose house arrest. The isolation gave him ample time to work on his theory, making him possibly the happiest prisoner of war in history. By the time the war ended and he was able to return to Belgrade, Milankovitch had completed a theory of the Earth's current climate, published in 1920 under the not exactly best-selling title *A Mathematical Theory of the Thermal Phenomena Produced by Solar Radiation.* That marked the beginning of a cascade of climate papers that emerged from Milankovitch's pen over the next two decades, culminating in 1941 with the publication of a massive work called *Canon of Insolation and the Ice Age Problem,* which laid out his complete theory of how orbital changes could cause ice ages. ("Insolation" refers to the total amount of solar energy that a given area of the Earth receives over a certain period, not to be confused with *insulation,* a product that lowers your home heating bills.)

There are at least three reasons why Milutin Milankovitch, and not James Croll, gets most of the credit for the orbital theory of the ice ages. First, he had better data than Croll on the amount of solar energy arriving at the top of the atmosphere, allowing him to calculate more accurately how solar heating changes with shifts in the Earth's orbit. Second, his theory took better account of how heat is moved around the Earth by winds and ocean currents. Third, he refined and expanded Croll's orbital parameters, ultimately coming up with three long-term cycles. The first is the hundred-thousand-year *eccentricity* cycle that Croll worked out. (That cycle, remember, is the shift in the Earth's orbit from more circular to more elliptical and back.) The second is the *obliquity* cycle, which measures changes in the tilt of the Earth's rotation axis. Today, the Earth's axis tilts by 23.5 degrees from vertical in relation to the plane of its orbit. Milankovitch showed that this tilt shifts between 22.5 and 24.5 degrees and back again on a cycle of forty-one thousand years. That may not seem like much, but the Earth's tilt causes the winter and summer seasons, and Milankovitch showed that even small changes in tilt, when combined with other cycles, can make a big difference in how cold or warm the seasons become—a big factor in determining whether or not snow will build up to make ice sheets. The third cycle is *precession.* Spin a top, and you will notice that its axis traces a slow circle. The Earth's axis does that too, taking twenty-three thousand years for its axis to trace a full circle. Today, the North Pole points at Polaris, making it the North Star. But the slow gyration of the Earth's axis means that eleven thousand years from now the North Pole will point at Vega, coming back to Polaris twelve thousand years after that.

Orbit, tilt, gyration. When Milankovitch overlaid these three cycles, he noticed that they created a regular pattern of change over time in how much solar energy the Earth absorbs at different latitudes. Generously crediting Croll's groundwork, Milankovitch proposed that ice ages come and go according to this pattern. Today, we know that he was right. The timing of the ice ages and the interglacial periods in between match Milankovitch's orbital cycles too closely to be coincidence. We honor the insight by calling these changes *Milankovitch cycles,* and you will find them amply described in every modern climatology and geology textbook. But when Milankovitch proposed his theory, he was tripped up by the same problem that had flummoxed Croll. Geologists still had no way to accurately date past ice ages and, thus, no way to confirm the timings that Milankovitch's theory predicted. When Milankovitch died in 1958, his theory remained little more than an intriguing speculation. But that began to change a decade later, thanks to one of the greatest expeditions in the history of ocean research: the Deep Sea Drilling Project, or DSDP. The secret to dating the ice ages, it turns out, lay not on land where the ice had once been, but miles down on the deep ocean floor.

Scoop up a handful of muck from the abyssal seabed and put it under a microscope and, more often than not, you will see numberless tiny shells of plankton: the skeletal remains of little animals and plants that drift in ocean surface waters. These shells accumulate by the trillions on the deep seabed as generations of plankton die and sink. Starting in 1968, the DSDP began drilling holes into the deep ocean floor worldwide to test the then revolutionary idea of seafloor spreading (see text box). At each drill site, ships bored into the seabed to bring up core samples: cylindrical tubes of sediment, four inches wide and often hundreds of feet long, containing layers that stretched back millions of years. As geologists cleaved open these core samples, they found abundant fossils of a type of plankton called *foraminifera,* or "forams" for short. Some foram shells look a bit like a bunch of tiny golf balls glued together (figure 7.9), while others look like tiny snail shells. (The resemblance to snails is incidental; forams are not true animals but single-celled protozoans.) Forams build their shells with calcium carbonate, a mineral that contains calcium, carbon, and oxygen, which they extract from seawater as they grow. The beauty of forams is that *the oxygen in their shells reveals the waxing and waning of the ice ages.* The details are a bit involved, but if you've made it this far you deserve a full explanation.

Oxygen atoms come in three forms, or isotopes: ^{16}O, ^{17}O, and ^{18}O. The lightest, ^{16}O, has two fewer neutrons than ^{18}O. Water (H_2O) molecules made of ^{16}O are thus lighter than those made of ^{18}O and evaporate more easily from the ocean to form clouds. This means that the rain and snow that fall from clouds have relatively more ^{16}O, and less ^{18}O, than ocean water. During ice ages, ^{16}O-rich snow builds up at the poles,

FIGURE 7.9. A planktonic foraminiferan of the genus *Globigerina.* Foram shells contain a chemical signature that reveals both the temperature of the ocean water in which they lived and the size of polar ice caps. By analyzing foram shells in sedimentary layers that have accumulated for millions of years on the deep seabed, we can reconstruct the history of the ice ages. The shell is about a half-millimeter across—the size of a tiny grain of sand. (Photograph by Hannes Grobe, Wikimedia Commons.)

leaving the oceans slightly enriched in ^{18}O. Forams that live in ^{18}O-rich ocean water—in other words, ice-age ocean water—have relatively more ^{18}O in their shells. Water temperature adds to this effect. The colder the ocean water, the relatively more ^{18}O forams add to their shells, regardless of the amount of ice at the poles. The two effects together mean that during ice ages, when ice caps are large and polar ocean waters cold, forams add more ^{18}O to their shells, whereas during interglacial periods, when ice caps are small and polar oceans warmer, forams add less ^{18}O. The ratio of ^{16}O to ^{18}O in fossil forams thus gives us a thermometer that tracks the waxing and waning of ice ages—a fabulous discovery pioneered by the famed geochemist Harold Urey.

A Spreading Seafloor?

The Deep Sea Drilling Project arose in the 1960s as a way to test the notion that the seafloor grows and spreads from mid-ocean ridges—a process fundamental to the (then new) theory of plate tectonics. Starting in 1968, the research vessel *Glomar Challenger* began taking core samples from the ocean floor on both sides of the Mid-Atlantic Ridge. These samples soon revealed three things about the ocean floor that confirmed seafloor spreading: The age of the volcanic basalt beneath the sediment layers increases steadily with distance from the mid-ocean ridge; the thickness of the sediment layers above the basalt increases with distance from the ridge; and polarity reversals of the Earth's magnetic field form a pattern in ocean-floor rocks that makes sense only if seafloor spreading occurs. In later years, the Deep Sea Drilling Project expanded into the Ocean Drilling Program, followed by the Integrated Ocean Drilling Program, which is active today.

But one problem remains: How can we figure out the ages of those deep-sea layers in which the foram fossils are found? You might think we could use those dating methods that I described earlier in the chapter, but no. Radiocarbon (^{14}C) isn't much help because it doesn't go back past seventy thousand years, and uranium–thorium and amino acid dating are no good because foram shells don't contain much of either. However, geologists during the 1970s discovered a clock that worked beautifully for the deep-sea layers in which foram fossils reside. It wasn't in the fossils, but in the layered sediments themselves. Studying lava flows in Hawaii and other places, geologists had found a curious thing: The direction of the Earth's magnetic field flips, apparently at random, every few thousand to few million years. They could tell because lava contains small magnetic minerals that line up, like tiny compasses, with the Earth's magnetic field as the lava solidifies. These mineral compasses showed that sometimes the Earth's magnetic field points north, like today, and at other times it flips and points south. Because lava rocks are readily dated using radioactive isotopes, geologists were able to piece together a picture of magnetic reversals over time. The layered sediments of the deep sea don't have any lava flows, but they do contain little magnetic minerals blown off the land during windstorms. As these mineral grains settle to the seabed, they line up with the prevailing magnetic field, just as they do in lava rock. Each deep-sea core thus contains a Morse code–like pattern of magnetic flip-flops, and by counting back through that pattern, we can tell the ages of particular layers.

With oxygen isotopes as the thermometer and magnetic flip-flops as the clock, geologists were finally in a position to either vindicate or reject the Croll–Milankovitch orbital theory of ice ages. The judgment came in a classic paper published in the journal *Science* in 1976. There, the authors—James Hays, John Imbrie, and Nick Shackleton—showed that the peaks and valleys of the ice age curve matched nearly perfectly with Milankovitch's orbital cycles. More than a hundred years after James Croll had first proposed the connection, we had proof: The Earth's orbit is the pacemaker of the ice ages.

FUTURE ICE AGES?

Will another ice age come? Maybe. Earth is still going through Milankovitch's cycles, and the basic configuration of the continents and the paths of the ocean currents that move heat around the planet aren't likely to change much in the next few million years (see text box). Interglacial periods of the past two million years have typically lasted eight thousand to twelve thousand years. Our present one began about twelve thousand years ago. All things being equal, we should expect ice sheets to begin bulldozing southward across Canada and northern Europe within a millennium. But all things are not equal. In the past century and a half, humans have altered the atmosphere more than any species that has ever existed, save one.* Carbon dioxide from burning fossil fuels, along with lowered absorption of carbon dioxide because of deforestation, means that CO_2 levels in the atmosphere are now approaching double what they would probably be if humans had never existed. CO_2 is a heat-trapping gas; the more of it there is in the air, the more the Earth heats up. Most climatologists forecast significant warming of the planet over coming decades because of excess CO_2. Virtually all the human history that matters to us—from the dawn of agriculture to the invention of the computer chip—has taken place during the latter half of the present interglacial period, when both the sea and the climate have been exceptionally stable. Today, millions of us live within a few

* News reports on climate change often remark that no creature has ever changed the Earth's atmosphere more than humans, but that's not correct. When cyanobacteria invented photosynthesis more than three billion years ago, they introduced oxygen gas (O_2) into a world that previously had none, changing the Earth's surface chemistry irrevocably and making life as we know it possible.

The Ice Cometh . . . But Not Often

You may have the impression from this chapter that ice ages have been common in the planet's past. In fact, the Earth has been ice free for most of its existence. No dinosaur ever saw an ice sheet. In part, this is because the configuration of landmasses and ocean currents has kept the Earth's poles relatively warm for much of the past, despite the fact that Milankovitch's cycles have always happened. Ocean currents are like gigantic heat conveyor belts, distributing the Sun's energy from the warm tropics to the cold poles. Only when the arrangement of landmasses and ocean currents allows the poles to grow sufficiently cold can Milankovitch cycles assert control.

Evidence from glacier-scraped rocks and glacier-bulldozed sediments suggests that the Earth has experienced ten to twelve prolonged glacial periods over its history, including the one we are in now. I say "in now" because, although we are presently experiencing warm interglacial climates, the Earth has been ice-dominated for most of the past two-plus million years, and we should expect another glacial period to come within a millennium or so (unless human-caused global warming postpones or cancels its arrival). Difficulties of age dating make it hard to determine how long the ice-dominated periods of the past have lasted. But over the span of geologic time, it's clear that ice-free periods have been far more common than icy periods. Ice ages probably comprise less than 5 percent of the Earth's 4.6-billion-year history.

vertical feet of sea level, and billions depend on established regimes of precipitation and temperature to grow our food. It's hard to say whether human-induced climate change will prevent the next ice age or just postpone it. But change is surely coming.

CALIFORNIA'S COASTAL WETLANDS: THE ICE AGE CONNECTION

White pelicans are spectacular birds, and when flocks of them swoop by on nine-foot wings to land in Southern California's coastal wetlands, everyone takes notice. Most of them come only briefly to the wetlands—Tijuana Estuary, Bolsa Chica Marsh, Mugu Lagoon, and others—before taking off again to merge onto the Pacific Flyway. Like an avian highway in the sky, the Pacific Flyway is the route followed by millions of shorebirds and waterfowl on their seasonal migrations between South America and Alaska and points in between. Southern California's coastal wetlands—variously called *lagoons, marshes,* or *estuaries*—form rest stops where a migrating bird can set down safely amid the sea of urbanity that now floods much of the coastal zone. Several hundred migratory species stop every year, joining for a time the resident grebes, loons, coots, stilts, avocets, whimbrels, sandpipers, egrets, herons, and ducks. One and all, they devour the lush fare of the wetlands, especially worms, molluscs, and fish. Anyone who watches hundreds of wetland birds feasting without letup comes to appreciate the stunning productivity of wetland ecosystems. But Southern California's coastal wetlands are more than dining halls. They are also nurseries. Dozens of fish species navigate from the open ocean to the wetlands to mate and lay their eggs. Dip a net into a wetland tidal channel and you may catch infant halibut, turbot, bass, mullet, and other open-ocean fish before they become open-ocean fish. The National Marine Fisheries Service estimates that three-quarters of all commercial fish, and perhaps 90 percent of fish caught for recreation, depend on wetland habitat.

Like the staircases of marine terraces we explored earlier in this chapter, Southern California's coastal wetlands are products of the ice ages. Trace each wetland inland, and you will usually find that it merges with one or more stream valleys. Visualize a stream flowing toward the sea, say, a million years ago (figure 7.10a). As polar ice forms, the sea drops, and the stream responds by cutting a valley down to this lower sea level (figure 7.10b). When the ice melts, the sea rises and floods the valley to form a deep-water bay (7.10c), which eventually silts in to sea level to form a wetland (7.10d). Another cycle of sea fall-and-rise repeats the pattern. If the stream doesn't scour away all the sediment that previously filled the valley, some may be left stuck to the valley walls (7.10e, f). If the land has been rising this whole time (which it usually has been in Southern California), then the former wetland floor may be stranded above the new wetland, forming a terrace (7.10g). Cut, flood, fill, repeat. Today's Southern California coastal wetlands are a momentary snapshot from a movie called *Earth: The Past Two Million Years,* brought to you by the cycle of the ice ages and the planet's orbital wobbles.

As I pointed out earlier, modern society has developed and expanded across the Earth during a period of exceptionally stable sea level that has persisted for seven millennia (figure 7.5b). If we limit our perspective to human history, we might conclude that such stability is normal. But the cycle of the ice ages tells a different story. For most of the past three million years, the sea has been either rising or falling by as much as several feet per century. What has this meant for Southern California's coastal wetlands, and what does it imply for their future? Practically by definition, coastal wetlands occur at sea level, and so they must migrate with the rise and fall of the sea. Coastal wetlands of the past have migrated for *miles* back and forth across the continental shelf, continually dispossessed and forced either landward or seaward by the rise and fall of the sea. Run your fingers across any Southern California nautical chart and find the sixty-fathom (360-foot) depth-contour; that's about where the shoreline was twenty thousand years ago, during the Last Glacial Maximum (figure 7.5b). Just landward of there, envision ancestral avocets poking in mud flats, and ancestral mullet leaping in tidal channels.

Good-hearted environmentalist friends of mine furrow their brows with worry as they ponder forecasts of rising sea levels. "What will become of our coastal wetlands as the sea rises with global warming?" they wonder. The answer, I think, is that the wetlands will migrate, as they have done, many times, with the rising (and falling) seas of the past. "But it's different now," my friends protest. "Now there's no room for the wetlands to migrate inland." They point out how the flat marshes today often butt up against steep valley walls. It looks as if there's no place for the wetlands to go. But those steep valley walls didn't just appear; they have been there through many cycles of sea rise and fall (figure 7.10). Wetlands are extremely good at two things: trapping sediment washed in by gullies, streams, and tides; and producing organic matter, primarily plant debris. If the sea rises, say, three feet in the next one hundred years, it's not as if our coastal wetlands will then be three feet underwater, with no more mudflats, tidal channels, and associated productive habitats. Hundreds of feet of sand, silt, and organic matter fill the deep valleys below the wetlands today. It accumulated in the valleys as the sea rose. As the sea continues to rise in the future, sedimentation will most likely keep pace, maintaining the marshes and mudflats much as they are today. Most projections forecast between two and six feet of sea rise in the next century.* Leaving aside the fact that coastal wetlands endure roughly six feet of sea rise and fall every day (i.e., the tides), let's put that forecast—two to six feet of rise over the next century—into perspective. Looking again at figure 7.5b, notice the steep part of the graph where the sea rose at its most rapid pace between about seven and fifteen thousand years ago. That translates to a rise of four to five feet per century—right within the range of future

* Based on *Sea-Level Rise for the Coasts of California, Oregon, and Washington: Past, Present, and Future,* by the National Research Council, Committee on Sea Level Rise in California, Oregon, and Washington (National Academies Press, 2012).

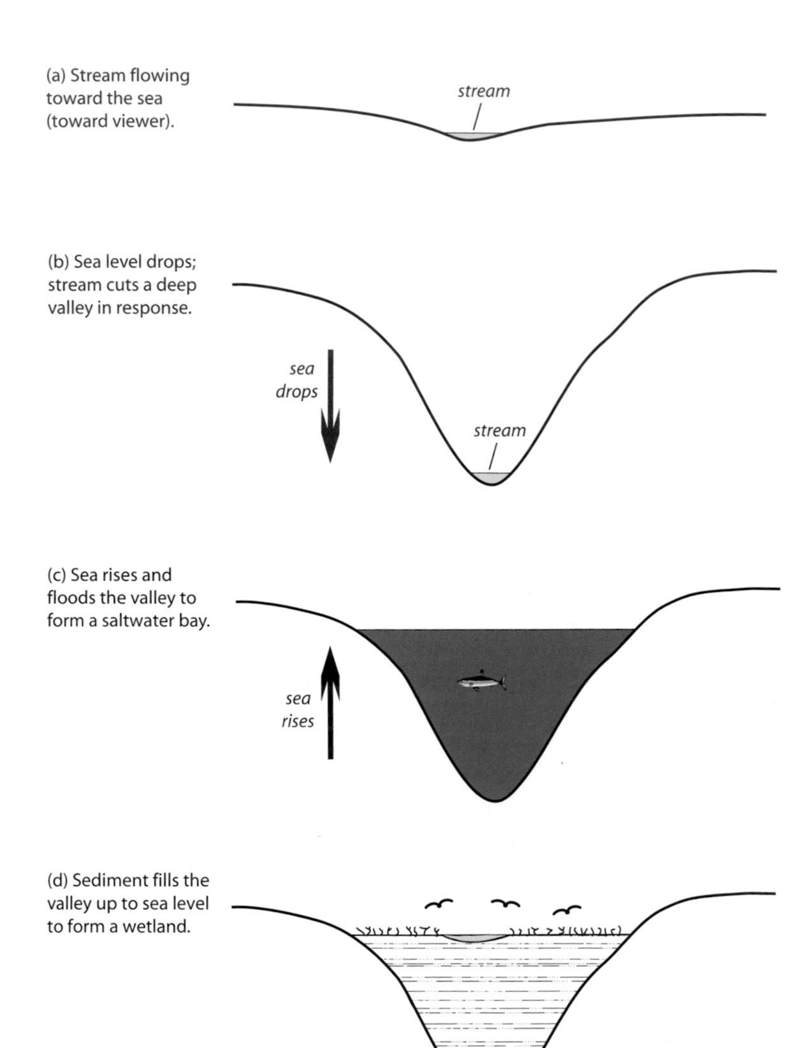
(a) Stream flowing toward the sea (toward viewer).
stream
(b) Sea level drops; stream cuts a deep valley in response.
sea drops
stream
(c) Sea rises and floods the valley to form a saltwater bay.
sea rises
(d) Sediment fills the valley up to sea level to form a wetland.

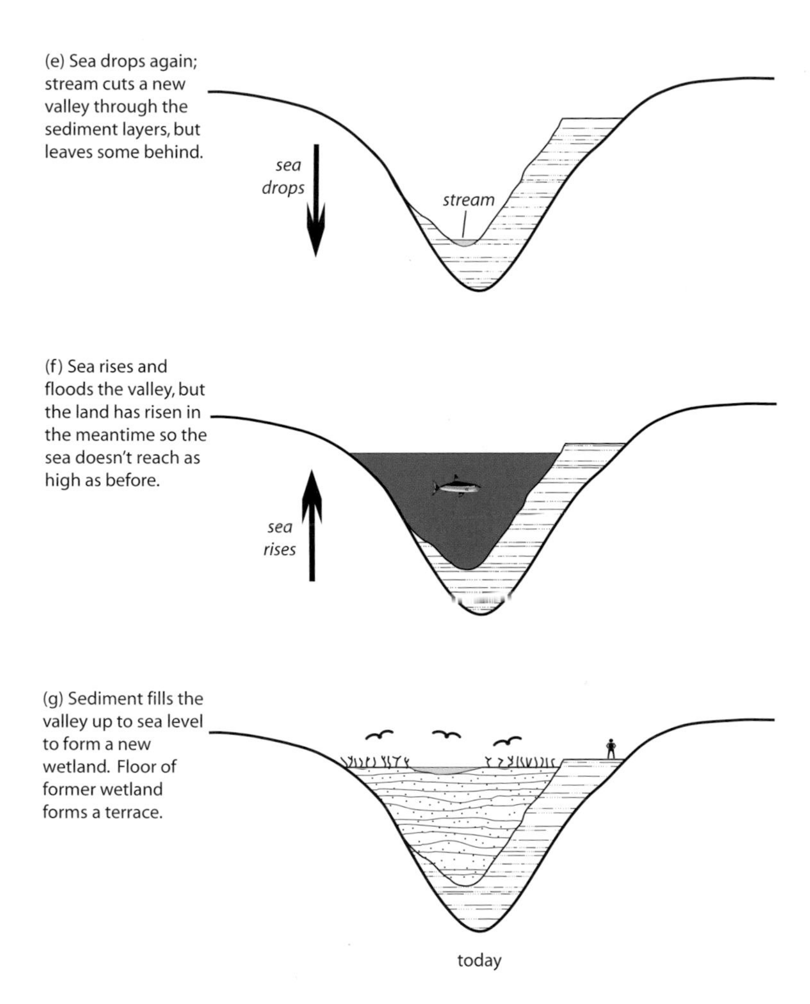

FIGURE 7.10. The fall and rise of the sea during glacial–interglacial cycles has caused streams to alternately cut and fill their valleys. The sea-drowned lower reaches of these stream valleys commonly form wetlands—variously called *lagoons*, *marshes*, or *estuaries*—along parts of Southern California's coast. See text for discussion.

forecasts. No evidence indicates the disappearance of wetland habitats, or the extinction of wetland species, during this period of rapid sea rise (or any prior period of sea rise, for that matter). Our coastal wetlands will likely be able to handle future sea-level rise.

The question is whether *we* can handle it. And that brings me to some final thoughts about surf, sand, stone, and our relationship with the coast.

Afterword

When you walk along a Southern California beach, a sense of timelessness can come to mind. The sand scrunches predictably underfoot, the coastal bluffs loom, seemingly unchanged since the last walk, and the sea brushes the shore with its same ceaseless rhythm. Yet most of us recognize the evanescence of the scene. Waves from a single storm may strip away much of that sand. Portions of that bluff may collapse without warning, continuing a natural retreat that dates back millennia. A large earthquake could elevate a portion of the coast several feet in an instant, stranding mussel beds above the tide line. And if we flip back through the last few pages of deep time—the past ten million years or so—the coastal scene, far from appearing permanent, looks like frenetic animation. The sea bobs up and down like a yo-yo. Violent earthquakes crackle without letup. Tsunamis wash ashore. Mountains lurch upward. Basins founder to abyssal depths while islands rise out of the sea. Old shorelines are jacked high into the air, and Southern California's geography undergoes a radical makeover as the Pacific Plate drags off large chunks of former North America toward the northwest.

But aside from the intellectual satisfaction of understanding these things, one could ask, "Does it matter?" Is there any reason to care, aside from curiosity? I think yes. We should care about the geologic and oceanographic processes that have shaped our world for practical reasons. An understanding of the past can, and should, inform decisions that we make today.

Take earthquakes and tsunamis. When it comes to these hazards, the past sends a clear message: More will come. The plate-tectonic rearrangement of Southern California (the story I began in chapter 1 and fleshed out in chapter 4) took millions upon millions of earthquakes. That rearrangement continues. We can neither prevent earthquakes nor predict them. But we can prepare. Earthquake codes for buildings and freeways in California are already among the strictest in the world, but if my polling of friends and students is any indication, many Californians remain underprepared. A simple guide is this: In the event of a major quake, be ready to survive for three days without electricity, gas, running water, or help. I keep two six-gallon jugs of water in my garage, plus several days' worth of canned food. If you need more water, remember that water-heater tanks and toilet tanks (the tanks, not the bowls!) hold several gallons of potable water. A first-aid kit with the capacity to treat severe cuts is essential, as are several working flashlights and fire extinguishers. Everyone in your household should know how to turn off the gas and electricity. (Hang a wrench permanently near the gas shutoff valve.) Keep slippers or sandals by your bed; big quakes can mean broken windows and mirrors, and the last thing you need is to have your feet cut up when you leap out of bed on a bad night. Think about getting a camping stove if you don't have one; it might make an emergency less unpleasant by providing hot water and food. (Your underprepared neighbors will love you.) Finally, many of us, if we're not at home, are either at work or commuting between home and work. So it's a good idea to have a basic earthquake kit (water and first aid) at your workplace and another in your car.

You've probably seen tsunami-warning signs posted in low-lying areas of the coast. An earthquake on an offshore fault could well deliver a tsunami to the mainland, either from direct fault displacement of ocean water or from an undersea landslide (chapter 2). The tsunami warning signs point you in the direction to go to get out of the hazard zone. To find out where the hazard zones are, you can download tsunami inundation maps for every part of the mainland coast.* The maps predict what areas will be hit by a maximum likely wave. The maps send a straightforward message: The tsunami hazard zone is mostly about elevation, not distance from the shoreline. A home perched on an

* Detailed tsunami inundation maps are available online (as of this writing) through the California Department of Conservation at www.quake.ca.gov/gmaps/WH/tsunamimaps.htm.

ocean-view lot thirty feet above the sea will probably be safe, whereas a road two miles inland that crosses a wetland a few feet above sea level will likely be flooded. In practical terms, if you feel an earthquake and find yourself near sea level, try to get to high ground as quickly as possible, even if you aren't near the beach.

A persistent lesson that coastal science teaches is that stability is an illusion, and when we invest in that illusion, we ask for trouble. For instance, when we crowd coastal bluff tops with ocean-view homes, we deny history. The bluffs look the way they do because of hundreds of feet of natural retreat (chapter 6). Imagine Southern California's bluff tops as great belts of parkland extending well inland from the bluff edge, with walking trails and bike paths instead of cramped private lots doomed to erosion, their fates merely postponed by cement. In this vision, the bluffs would continue to erode naturally, nourishing the beaches with fresh sand and requiring only the occasional rerouting of paths and trails. Instead, we have panicky investments in pricey seawalls that only put off the inevitable while starving our beaches of needed sand. Some may sneer at this as pie-in-the-sky thinking. But if California citizens decades ago had known as much as we do today about bluff erosion rates, sea-level changes, and beach sand budgets, different decisions might well have been made about bluff-top development—at least in some areas.

Another illusion is the stability of the sea. As I pointed out in chapters 6 and 7, the stable sea level to which human society has become accustomed in recent millennia is not typical of the past. Pick any century at random from the past few million years, and odds are you would witness the sea rising or falling, not holding steady. We presently face a probable sea rise of two to six feet over the next century as the Earth's climate warms. (If large portions of the ice sheets on either Greenland or Antarctica break up, as some experts fear, we could be in for quite a bit more than that.) The obvious implications are gradual inundation of harbors and low-lying communities and accelerated erosion of coastal bluffs. We should add to this more powerful storm waves created by a warmer Pacific Ocean. (In fact, bigger Pacific waves seem to be here already; measurements by wave buoys in recent decades show an increase in the frequency of large waves in the Southern California Bight; see figure A.1). These developments all point to a looming shore-erosion crisis. How should we handle it? Perhaps through a combination of managed retreat where possible, and sustained beach replenishment where retreat is not an option. But the replenishment will have to be on a heretofore-unprecedented scale. We will need to pump enough

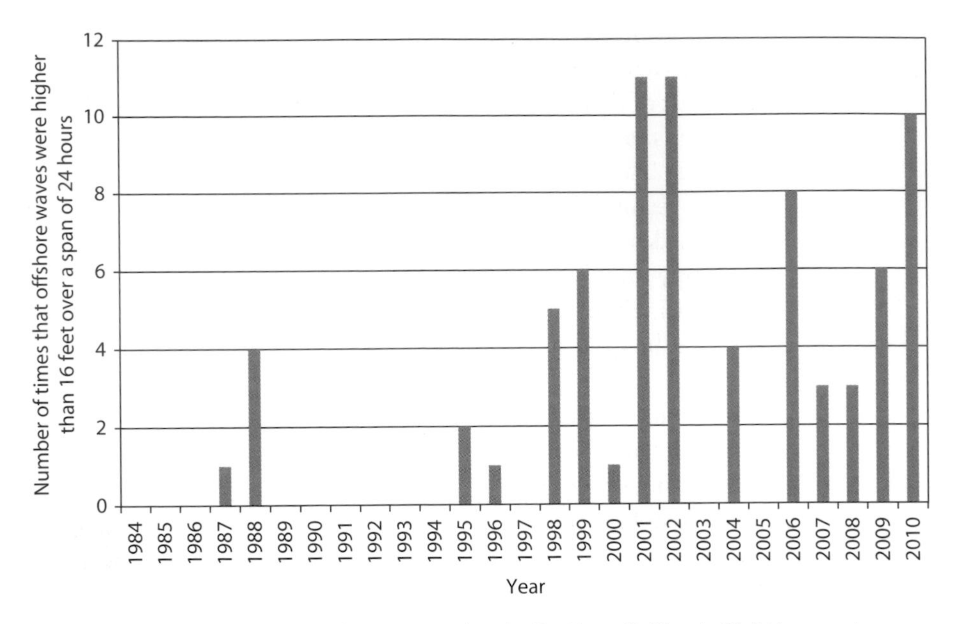

FIGURE A.1. Increasing heights of waves entering the Southern California Bight in recent decades, based on data from offshore buoys. Although the time span may be too limited to draw firm conclusions, the apparent trend toward larger waves is striking. (From Russell and Griggs 2012: fig. 3.10.)

sand onto the beaches to make up for increased losses from storm erosion, to compensate for ongoing sand starvation from dams and seawalls, and to keep up with the rising sea. I'm reminded of the scene in *Alice in Wonderland* where the Red Queen explains to Alice, "It takes all the running you can do to keep in the same place."

The problem of rising seas connects to the broader issue of climate change. As our trip through the ice ages (chapter 7) showed, there's nothing unusual about climate change. Natural climate warming brought the Earth out of the last ice age. But human carbon emissions are accelerating this natural warming and pushing it beyond where it would normally go. One likely ramification for Southern California is a reduced water supply. Most models of future atmospheric circulation predict that rainfall and snow packs will shrink across much of the southwestern United States in coming decades. All of the fresh water we use in Southern California ultimately rises from the ocean. As the sea evaporates, water molecules (though not salt) rise into the air, condensing into cloud moisture that eventually falls as rain or snow, mostly in mountain ranges. Our water supply depends largely on moisture that falls in the Rockies and the Sierra Nevada, with the water delivered here through a vast network of aqueducts. (Only a few areas of Southern California get most of their water from local wells and reservoirs.) If the supply from these sources shrinks as predicted, we will need to figure out other options. Three come to mind: conservation, recycling, and desalination. The technologies to do all three already exist, and we need only look to other desert cultures, particularly Israel, to see them implemented successfully on a large scale.

Although municipalities can do much to conserve water, agricultural reform could yield bigger savings. Agriculture uses most of California's fresh water. While recent droughts have motivated farmers to adopt some conservation methods, much of the water is still used for inefficient flood-irrigation, in which whole fields are flooded, with a high portion of the water evaporating or seeping into the ground unused. Israeli agriculture, by contrast, employs efficient drip irrigation to deliver water directly to plant roots on computer-regulated schedules, along with abundant greenhouses that extend growing seasons while conserving water. Moreover, Israelis have steadily replaced fresh water with treated urban wastewater to raise crops. Any "ick" reaction one might feel at using wastewater to grow crops has no rational basis. Technologies already exist to clean sewage water to a potable level. (Every water molecule on the planet has been around for several billion years, and over that span some of them have surely cycled through places that we would rather not think about.) Israelis treat their wastewater to near potability before it enters the irrigation system, and they now grow more than half their crops with recycled wastewater. By contrast, coastal Southern California cities pump millions of gallons of treated wastewater into ocean outfall pipes every day. Only a fraction is recycled, mostly for noncrop irrigation such as golf courses and public landscaping.

Along with conservation and recycling, we can increase our fresh water supply through desalination, which effectively means that we bypass Nature's system for delivering fresh water from the ocean through evaporation and precipitation, and do the job ourselves. At this writing, a fifty-million-gallon-per-day desalination plant is close to completion in Carlsbad in northern San Diego County. It will be the largest desalination plant in the Western Hemisphere and should soon supply about 7 percent of San Diego's municipal water needs. Desalinated water is expensive. Reverse osmosis, the main large-scale desalination method in use today, works by pushing seawater at high pressure through tiny filters that separate salt from water. The method uses a lot of electrical power, at a cost to both consumers and the environment in the form of carbon emissions used to generate electricity. (Curiously, a less significant environmental problem—the effects of higher salinity on marine life near desalination plants—often attracts more concern. If the salty outflow from the plants is diluted with sufficient seawater before it reenters the ocean, the rise in saltiness near the plants can be kept well within the natural tolerance range of marine species.) Desalination technology continues to improve; a gallon of desalinated water today uses only a quarter of the electricity it did three decades ago, thanks to improved pumps, filters, and energy-recovery systems. The Pacific Ocean is Southern California's only guaranteed source of water. No matter how bad the drought, the sea will always be there.

Do you remember those immigrant Mexican pebbles on San Miguel Island whose story I told in chapter 1? You can find those same distinctive pebbles in canyon walls and bluffs near Black's Beach on the San Diego coast. I take my students to these outcrops. They hold the pebbles (some as big as baseballs) in their hands while I tell them the story of how Southern California came to be. (You can flip back to figure 1.4 to remind yourself of the pebble story.)

One afternoon some years ago, after finishing the field trip and sending the students off, I went for a swim. Swells from a big North Pacific storm were rising up and peeling off in the classic lefts for which Black's Beach is famous. After body-surfing some smaller waves, I swam farther out to catch some bigger ones. Before I knew it, a bruiser of a wave picked me up high onto its crest. As it toppled forward in a great curl, it sent me flailing "over the falls"—a surfing phrase that describes being caught on the waterfall-like face of a plunging breaker. The cascade drove me straight down to the bottom—smack!—and then scrubbed the sandy seabed with me as it tumbled me toward the beach. Held underwater too long for comfort, I clawed my way back to shore. Distracted by a bleeding elbow, I then stubbed my toe against a large stone as I walked up the beach. "Surf, sand, and stone," I pondered as I limped back to my car, "they all took turns beating me up just now." As I drove home, an idea for a book took hold in my mind.

Acknowledgments

Luck and circumstance have allowed me to earn a living doing what I like best: learning the science of the outdoor world and teaching it to college students. I suppose it all began with my mother, who, decades ago, watched with some alarm as my older siblings sallied forth from college with degrees in the humanities only to flounder in marginal employment. Seeing me headed in the same direction (I was a history major), she said, "You've always liked science, why don't you try that? You might get a *job!*" Dubious, I enrolled in a college geology course. Early in the course, the professor showed us pictures of scruffy geologists floating on rafts down the Grand Canyon, drinking beer and looking at rocks. "This is what geologists do," he said. As a recruitment strategy for new majors, it remains the best I have ever seen. Research for a PhD followed my bachelor's degree, taking me, a former New Englander, to the glorious American West. My ramblings eventually landed me on the Southern California coast near San Diego, where I have now lived for close to two decades.

I thank my colleagues at Mira Costa College for supporting faculty sabbaticals as part of a robust program of professional development. Without that support, this book, and my previous efforts, might never have come to fruition. I thank California's taxpayers for supporting our state's community college system, perhaps the finest in the nation, where, for not much money, students can enter higher education in small classes staffed by men and women whose number-one priority is to teach well. Enrolling in a California community college may be one of the best educational decisions any student can make.

I thank the staff at the University of California Press. Blake Edgar (who helped pilot my previous book *Rough-Hewn Land* to completion) guided this effort skillfully from proposal to final product and read everything with a keen editorial eye. Supportive reviews of the finished manuscript came from professors Gary Griggs (University of California at Santa Cruz) and Patrick Abbott

(San Diego State University). Merrik Bush-Pirkle helped with coordinating the art program, securing the cover rights, and other logistics. Kate Hoffman's suggestions improved the book's flow and organization. Richard Earles's copyediting clarified the writing and fixed a number of errors. My brother Malcolm has been my faithful critic and editor for ten years and three books. His pencil is an apex predator: merciless, inescapable, and indispensable, thinning the herd of clichés, verbiage, and those insidious, no-account, outrageous, bewhiskered adjectives.

Finally, I thank my wife and sweetheart, Susan, who has brought me happiness and the satisfaction of a shared passion for both science and the outdoors. I finally got a chance to dedicate a book to you, my dear—it's about time.

APPENDIX

Seeing for Yourself

This appendix will guide you to many of the key sites that informed my research for *Surf, Sand, and Stone*. The latitude–longitude coordinates, typed into Google Earth or Google Maps, will give you the location of the outcrops, viewpoints, or other features described. Some of the sites are not explicitly mentioned in the book, but I have included them here because they illustrate concepts or evidence that informed portions of the book. Sites along the mainland are listed from south near San Diego to north near Point Conception, followed by selected offshore sites. Each entry includes a brief description and a reference to the chapter(s) and figure(s) to which it connects.

MAINLAND SITES

Tijuana River Estuary Reserve, Imperial Beach, San Diego Co.

Connection: Chapter 7, section on coastal wetlands and sea-level changes.

Where: 32.57483°, –117.12633° for the Visitor's Center at 301 Caspian Way in Imperial Beach, where you will find displays, trail maps, and other information.

What's There: The estuary is one of Southern California's largest coastal wetlands. It formed as the Tijuana River repeatedly cut and filled its valley in response to the falls and rises of the sea during the ice ages, as portrayed schematically in figure 7.10. Birdwatchers may want to bring binoculars.

Mount Soledad, La Jolla, San Diego Co.

Connections: Chapter 3, section on earthquakes along the Rose Canyon fault; chapters 5 and 6, sections on La Jolla Canyon; chapter, 7 section on marine terraces.

Where: 32.84011°, −117.24468° marks a large parking area by the mountaintop monument at the end of Soledad Road in La Jolla.

What's There: Panoramic views from Mount Soledad illustrate many features of San Diego geology. Look northwest to the beach south of the pier at Scripps Institution of Oceanography. If you could drain the ocean, you would be looking straight down the axis of La Jolla Canyon. Looking north and northeast, notice the broad mesas, heavily built over and bisected by stream valleys. The mesas are marine terraces—flat surfaces cut by waves at former sea levels—and they rise gradually in elevation and become older inland. (Figure 7.4 shows the pattern with great vertical exaggeration.) For instance, the University of California campus about two miles directly to your north is built upon a marine terrace that is about 800,000 years old, whereas the Miramar Naval Air Station, whose runways you can see about six miles to your northeast, lies atop one that is a bit higher and about 930,000 years old. Extend your gaze east beyond the air station to where the mountains rise steeply from the last of the mesas. That's where ocean waves were breaking about 1,290,000 years ago. Closer in, and using figure 3.6 as a guide, let your eye track the trace of the Rose Canyon fault, which sweeps around the base of Mt. Soledad in a broad bend to the northwest. This is a classic restraining bend in a fault (see figure 2.4), and the squeezing forces along the bend have pushed up Mt. Soledad. The marine terrace that forms the top of Mt. Soledad eight hundred feet above the sea is the same age as the one that underlies the University of California campus about four hundred feet lower. The difference in elevation reflects the ongoing rise of Mt. Soledad along the Rose Canyon fault. For more about the views from this location, see Abbott (1999: 212–214).

Scripps Coastal Reserve, La Jolla, San Diego Co.

Connections: Chapter 5, section on the surfing waves at Black's Beach; chapter 6, section on erosion of coastal bluffs; chapter 7, section on marine terraces.

Where: 32.87586°, −117.24603° marks the reserve entrance off La Jolla Farms Road. Park nearby and walk west one-fifth of a mile across the mesa to the bluff-top overlook.

What's There: The nearly flat mesa surface here is a marine terrace, originally cut by waves at sea level and since uplifted to its present elevation of 360 feet. Looking south toward the stubby peninsula of La Jolla and to Mt. Soledad, notice how the land looks like a staircase in profile. Each step is a marine terrace, and each higher terrace is older than the one below, as portrayed (with great vertical exaggeration) in figure 7.4. If you could drain the ocean here, you would be looking down into Scripps Canyon, the head of which begins directly offshore of Black's Canyon, the big canyon immediately to your north (see description below). If you are here when long-period swells (sixteen seconds or more) are arriving, you will likely see the refraction effects of the canyon curving the waves, as portrayed in figure 5.8. The bluffs in this section of the San Diego coast retreat about three to four inches per year, on average (although the erosion is highly episodic). Assuming that that rate has been typical of the last few millennia, one can roughly calculate where the edge of the bluff stood at various times in human history, as portrayed in figure 6.3.

Black's Canyon, La Jolla, San Diego Co.

Connections: Chapter 1, section on the immigrant Mexican pebbles and the plate-tectonic evolution of Southern California.

Where: 32.88004°, −117.24700° along La Jolla Farms Road marks a gated entrance to a road (pedestrian access only) that leads downhill to Black's Beach. Walk down the road about one-third of a mile, passing through a large S-shaped switchback, to an outcrop on the right (north) that contains abundant large, rounded pebbles at 32.87785°, −117.24973°, several hundred yards uphill from the beach.

What's There: Notice the common purple-maroon pebbles dotted with white crystals. These pebbles are identical in appearance and chemistry to volcanic lava beds in Sonora, Mexico, and to pebbles on the Northern Channel Islands (see figure 1.1; also the description for San Miguel Island below). The far-flung distribution of the pebbles reflects the sidling movements of the Pacific Plate northwest past the North American Plate. Those movements dismembered the pebbly deposits of the extinct Ballena River to lead to their present distribution, as shown in figure 1.4. Standing back from the outcrop, notice the broad, scoop-like shapes of the bottom edges of the pebble-bearing deposits. The pebbles lie in large channels cut by the Ballena River as it carried the pebbles from the mountains to the coast and deposited them in a delta.

San Elijo Lagoon, Cardiff, San Diego Co.

Connection: Chapter 7, sections on marine terraces and coastal wetlands.

Where: 33.01338°, −117.27404° for the Nature Center at 2710 Manchester Avenue in Cardiff.

What's There: The lagoon is one of several coastal estuaries between Oceanside and La Jolla. Each lagoon formed as local streams cut and filled their valleys in response to the falls and rises of the sea during the ice ages, as portrayed schematically in figure 7.10. If you follow each lagoon inland, you will come to one or more stream valleys. Looking south across the lagoon to the far side, notice the flat, step-like surfaces occupied by houses. These are marine terraces, cut by waves at former sea levels, and you can see that they rise in elevation inland. The youngest terrace is the lowest one, closest to the sea, and the oldest is the highest one, farthest from the sea.

Oceanside Harbor Breakwater, Oceanside, San Diego Co.

Connection: Chapter 6, section on how human shore structures interfere with longshore drift.

Where: 33.20935°, –117.40562° at the entrance to Oceanside Harbor.

What's There: Visiting the breakwater from land is difficult, and a better perspective can be had from the air via Google Earth. The breakwater angles into the sea on the north side of Oceanside Harbor, protecting the harbor from west and northwest swells but also blocking the natural southward longshore drift of sand. A view from the air shows that the beaches north of the breakwater are much larger than those to the south. Blocking of sand by the breakwater has created this imbalance. The south-drifting sand eventually works its way around the end of the breakwater and piles up at the harbor entrance. The sand needs to be dredged regularly to prevent blockage of the harbor mouth.

Trestles Beach at San Mateo Point, San Clemente, Near the San Diego–Orange County Line

Connection: Chapter 5, section on Trestles surfing.

Where: Exit Interstate 5 at Cristianitos Road, turn east, and park as soon as possible. About 150 yards east of the interstate, on the south side of Cristianitos Road at 33.39615°, –117.59156°, you'll find access to a foot and bike trail that leads downhill toward an old road along San Mateo Creek, passes underneath the interstate, and reaches the train tracks and the beach in three-quarters of a mile.

What's There: Where the trail reaches Trestles beach, the Cottons surf site is just to the north, and all other sites are to the south. Walking south a half-mile will take you past the mouth of San Mateo Creek to Lowers (33.382300°, –117.58871°), a bulging point in the beach where waves wrap tightly toward a shallow underwater ridge. Lowers is one of the most popular surf sites at Trestles and attracts some of the top surfing talent.

Dana Point, Dana Point Harbor, San Clemente, Orange Co.

Connection: Chapter 4, section on the tectonic evolution of the Continental Borderland, in particular the erosion of ancient highlands composed of Catalina Schist to form the bouldery deposits of the San Onofre Breccia.

Where: From the parking area by the Ocean Institute at the end of Dana Point Harbor Drive (33.46143°, –117.70704°), walk west past the buildings to the rocky beach just north of the Dana Point Harbor breakwater.

What's There: Walking from the parking lot west toward the beach and the breakwater, notice to your right (north) the east-tilting, boulder-rich layers in the cliffs. A highland source for these boulders once lay to the west, somewhere out where the ocean is today. This is a deposit of the San Onofre Breccia—a boulder-filled rock formation, common throughout Southern California, derived from erosion of now vanished highlands composed of another rock formation known as the Catalina Schist. (For more on the Catalina Schist, see the description for Two Harbors on Catalina Island below.) About fifteen to twenty million years ago, when the tectonic stretching and breakup of the Continental Borderland was in its early stages (see the first two time-panels in figure 4.8), highlands composed of Catalina Schist arose across what is now mostly open ocean to the west. As these highlands eroded, they shed copious boulders off their flanks, forming the San Onofre Breccia. Walk on west to the beach and north along the high cliffs. The San Onofre Breccia is gloriously displayed, with angular boulders as big as basketballs or bigger. The size and angularity of the boulders tells us that the highlands that shed the boulders could not have been more than a few miles to the west of here. Picture a large mountain or island a mile or two west of where you are standing, shedding boulders of Catalina Schist off its flanks into the Miocene sea. As the Continental Borderland continued to stretch and evolve toward its current configuration (see the last two time-panels of figure 4.8), these highland sources foundered below the sea, leaving only the bouldery deposits of the San Onofre Breccia to testify to their existence. For more, see Sharp and Glasner (1993: chapter 4) and Abbott (1999: 160–165).

Aliso Beach County Park, Laguna Beach, Orange Co.

Connection: Chapter 4, section on the tectonic evolution of the Continental Borderland, in particular the erosion of ancient highlands composed of Catalina Schist to form the bouldery deposits of the San Onofre Breccia.

Where: Aliso Beach County Park along Highway 101 in Laguna Beach. Parking area at 33.51038°, −117.75235°. From the parking area, walk south two-tenths of a mile along the beach to the bouldery outcrops of San Onofre Breccia that protrude onto the beach. Immediately beyond, you'll find yet more exposures of breccia in the sea cliffs.

What's There: These outcrops are excellent samples of the San Onofre Breccia, derived from erosion of the Catalina Schist. See the information given for Dana Point above, and for Two Harbors on Catalina Island below. For more, see Sharp and Glasner (1993: chapter 4) and Abbott (1999: 160–165).

Crystal Cove State Park, Laguna Beach, Orange Co.

Connection: Chapter 7, section on marine terraces.

Where: Crystal Cove State Park campground immediately east of Highway 101 in Laguna Beach, 33.56214°, −117.82288°.

What's There: Views north and south illustrate well the first flat "step" in the "staircase" of marine terraces common along much of the Southern California coast. The flat surface on which the campground sits is a marine terrace, presently one hundred feet above the sea. You can see the same flat terrace surface at about the same elevation both to the north and to the south. Waves cut this broad terrace at sea level more than a hundred thousand years ago. Since then, tectonic forces have jacked the land upward so that the terrace now sits high above the sea.

Los Angeles—Long Beach Seaport Complex, Long Beach, Los Angeles Co.

Connection: Chapter 2 on tsunami hazards.

Where: The seaport complex is immense, so taking it in means finding a high perch. The upper deck of the Queen Mary is one option (1126 Queens Highway, Long Beach, CA 90802: 33.75236°, −118.18832°). You'll also get good views from the high bridges on Highway 47 (Seaside Freeway) that cross the freighter channels on the east (33.76461°, −118.22120°) and west (33.74952°, −118.27153°) sides of Terminal Island, although traffic is usually heavy and there is no stopping on the bridges.

What's There: The entire ten-square-mile seaport complex lies just a few feet above sea level. Late-nineteenth-century maps show this area as a vast marsh called Wilmington Lagoon. A large tsunami, triggered perhaps by an undersea landslide in the Continental Borderland, would likely flood the entire seaport complex, causing billions of dollars in damage and crippling the import–export economy for some time.

Sunken City, Point Fermin, San Pedro, Los Angeles Co.

Connection: Chapter 6, section on coastal erosion at Sunken City, Point Fermin.

Where: A viewing trail runs along a fence that blocks off the area destroyed by land sliding. The trail is accessible from the south end of Pacific Avenue (33.70673°, −118.28798°) or the south end of Carolina Street (33.70640°, −118.29067°) in southern San Pedro.

What's There: Sunken City, the wrecked remains of a community destroyed by slow-moving landslides, represents a dramatic example of coastal erosion (figure 6.1). You will see displaced and broken roads, pipes, and house foundations. (The houses themselves have long since been removed.) In areas of exposed bedrock near the cliff edge, you may see sedimentary layers that tilt toward the sea. Some of these layers grow very soft and weak when wet, so as waves undercut the cliff below, the layers slide toward the sea.

Point Vicente, Rancho Palos Verdes, Los Angeles Co.

Connections: Chapter 7, section on marine terraces; chapter 2 on tsunamis.

Where: Point Vicente Interpretive Center at 31501 Palos Verdes Drive West, Rancho Palos Verdes, CA 90275 (33.74476°, −118.41151°). Walk out to the bluff top for panoramic views and, in the right seasons, gray whale watching. Exhibits inside the center highlight the cultural and natural history of the Palos Verdes Peninsula.

What's There: The Palos Verdes Peninsula exhibits some of the best marine terraces in Southern California. Looking toward the Point Vicente Lighthouse and using figure 7.6 as a guide, you can see that you are standing on the lowest, youngest marine terrace in the area. This terrace, cut by waves during a highstand of sea level about eighty thousand years ago, forms a broad platform that wraps nearly unbroken around the entire peninsula. Looking at the profile of the hills, you'll notice several other flat steps at higher elevations. These are older terraces cut during earlier highstands of sea level. Uplift of the Palos Verdes Peninsula along faults has raised all the terraces to elevations much higher than when they formed. Figure 7.7 illustrates how such staircases of marine terraces come to be. On a different topic, now look south and southwest across the sea toward Catalina Island. If you could drain the ocean, you would see that you are just one mile from the lip of the San Pedro Escarpment (see figure 2.7). Here, the seabed plunges rapidly to a half-mile deep, forming one of the steepest underwater slopes in the Continental Borderland. A large undersea landslide here could trigger a tsunami that would likely devastate low-lying parts of the coastline and the LA–Long Beach seaport, although the wave would not likely rise even close to the 130-foot-high terrace surface on which you stand.

Redondo Beach/King Harbor Marina, Redondo Beach, Los Angeles Co.

Connections: Chapter 5, section on surfing waves created by Redondo Canyon; chapter 6, section on how human shore structures interfere with longshore drift.

Where: The breakwater angles west from the beach at 33.85055°, −118.39945° in Redondo Beach just north of King Harbor Marina. Park near Yacht Club Way by the marina and walk to the beach just north of the breakwater.

What's There: The buildings immediately north of the breakwater block your view of how large the beach is here; if you walk past the buildings, you'll see that you're standing on a beach six hundred feet wide, one of the largest in Southern California. The breakwater blocks the natural southward longshore drift of sand, making the beach extra-wide. By contrast, the beaches south of the marina are much smaller because they are starved of sand by the breakwater. Walk west along the breakwater to check out the surfing action. If you could drain the ocean here, you would be looking nearly straight down the axis of ten-mile-long Redondo Canyon. On a large, long-period west swell, the refraction effects of the canyon cause the waves to bend north away from the canyon and sling monstrous surf toward the breakwater and beyond.

Malibu Point, Malibu, Los Angeles Co.

Connection: Chapter 5, section on surfing waves at Malibu.

Where: The main parking area for Malibu Lagoon State Beach is just south of coastal Highway 1 where it intersects Cross Creek Road in Malibu at 34.03416°, −118.68493°. A walking trail winds for a third of a mile past the lagoon to the beach at Malibu Point.

What's There: The first official World Surfing Reserve is at its best during long-period south and southwest swells, which are most common during summer, although you'll find decent surf here most of the year. Floods down Malibu Canyon have swept sand and rocks into the sea, pushing the shoreline south to make the bulge of Malibu Point and the lagoon behind it. South swells wrap to the rock-strewn shallows around the point to create classic, long-traveled right breaks (see figure 5.12).

Rindge Dam, Malibu Canyon, Santa Monica Mountains, Los Angeles Co.

Connection: Chapter 6, section on how river dams block beach sand.

Where: From where Malibu Canyon Road intersects coastal Highway 1 about one mile west of Malibu Point, drive north on Malibu Canyon Road for 2.6 miles and pull over on the right (north) side at 34.06350°, −118.69805°, just as the road straightens out after making a broad bend to the west. Follow the short dirt paths away from the highway to get an overlook down to Rindge Dam in the bottom of Malibu Canyon.

What's There: The reservoir behind the now decrepit dam, built in 1926, had filled in with sand by the 1950s and continues to retain vast amounts of sand that might otherwise widen the beaches downstream. Sand retention behind hundreds of dams has led to chronic sand starvation for many California beaches.

Malibu Canyon and the Santa Monica Mountains, Los Angeles Co.

Connections: Chapter 1, section on the plate-tectonic evolution of Southern California; chapter 3, section on earthquakes in the Big Squeeze.

Where: Continue north along Malibu Canyon Road from the Rindge Dam overlook described above. The road crosses the axis of the Santa Monica Mountains through steep-walled Malibu Canyon. Exactly two miles from the Rindge Dam overlook (or 4.6 miles from Highway 1 at Malibu), turn right (east) onto Piuma Road (intersection 34.08213°, −118.70449°) and climb steeply uphill. At 3.0 miles from the intersection, immediately after the road curves sharply left around a big switchback, turn left into a dirt overlook area at 34.07395°, −118.69773°. This is the first of two overlooks that take in Malibu Canyon and the upland drainage basin of Malibu Creek. To get to the second overlook, continue east uphill for another 0.5 miles to a narrow pullout on the right at 34.06941°, −118.69140°.

What's There: The Santa Monica Mountains are some of the youngest in the United States. They began rising in earnest about five million years ago and are still squeezing upward. The mountains lie in the tectonic region known as the Big Squeeze, where the Big Bend in the San Andreas fault causes the Pacific and North American tectonic plates to mash into each other (figure 1.3), resulting in geologically rapid uplift and a high risk of earthquakes (figure 3.2). Evidence for the youth and recent uplift of the mountains is well displayed from these two overlooks. The precipitous slopes and steep canyon walls speak of recent uplift; if the mountains were older, erosion would have smoothed the slopes. The uplift is so recent that Malibu Creek has had no time to widen Malibu Canyon; it has only been able to cut downward apace with uplift to create a narrow defile through the mountains. Notice the tilted rock layers exposed in the mountains and the canyon walls. These layers, only a few million years old, were originally laid down flat, like blankets on a bed. Their akimbo orientations testify to geologically recent uplift. From the first overlook, your view north takes in most of the upper drainage basin of Malibu Creek. The tributaries of Malibu Creek arise in the broad uplands to the north and converge below you, where the creek cuts southward through the high east–west axis of the Santa

Monica Mountains. It is exceedingly strange for stream tributaries to converge *toward* a large mountain. Normally tributaries arise *on* a mountain and converge downhill *away* from it. What is going on to create the situation here? The likely answer is that Malibu Creek and its tributaries flowed to the sea *before the Santa Monica Mountains rose.* Visualize Malibu Creek flowing south to the coast with no Santa Monica Mountains in the way. As the mountains arched upward, the creek sliced downward apace with uplift to cut Malibu Canyon.

Channel Islands Harbor Breakwater, Oxnard, Ventura Co.

Connection: Chapter 6, section on how human shore structures interfere with longshore drift.

Where: 34.156900°, –119.23214° marks the breakwater offshore of the entrance to Channel Islands Harbor near Oxnard.

What's There: Visiting the breakwater in person is perhaps less instructive than a perspective from the air via Google Earth. The 2,400-foot-long breakwater lies 1,000 feet seaward of the harbor entrance. Typically, a gigantic bulge of sand protrudes from the beach on the north side of the harbor entrance behind the breakwater (see figure 6.8). Net longshore drift of beach sand is to the south in Southern California. The breakwater, by damping the power of waves to move the sand south, causes south-drifting sand to accumulate here. The sand needs to be dredged regularly to prevent blockage of the harbor mouth. The sand comes mostly from the mouth of the Santa Clara River five miles to the north. The Santa Clara is one of the least-dammed rivers in Southern California, so it can still deliver substantial sand to the beach south of the river mouth.

Channel Islands National Park Visitor's Center, Ventura, Ventura Co.

Connection: Features of the Channel Islands discussed in various parts of the book.

Where: Located in Ventura Harbor at 1901 Spinnaker Drive, Ventura, CA 93001 (34.248300°, –119.26667°).

What's There: Exhibits on the geology, biology, and human history of each of the Channel Islands.

Ventura Hills, Ojai Freeway, Ventura Co.

Connection: Chapter 3, section on oil formation in Southern California.

Where: Along the Ojai Freeway (Highway 33) north from its intersection with Highway 101 in Ventura, particularly within the two-mile section between two and four miles north of Highway 101, centered at 34.31813°, –119.29269°.

What's There: The economic significance of this area is apparent as you approach from the south. The hills are dotted with oil derricks and pump jacks bobbing like gigantic locusts. The two-mile stretch of highway referred to above crosses the axis of the Ventura Anticline. An anticline is an upward arch of rock layers, analogous to bending a paperback book so that it folds upward in the middle. Anticlines form excellent traps for oil underground. The Ventura Anticline is one of the most productive in Southern California. Its axis runs east–west underground for about fifteen miles, and Highway 33 cuts perpendicularly across its center. If you check this area on Google Earth, you will see the anticline clearly marked by the many twisting access roads that lead to scores of pump jacks. The tectonic squeezing of this region has produced many anticlines, including throughout large parts of the Santa Monica and Santa Ynez mountains. For a more detailed guide to the Ventura Anticline, see Sharp and Glasner (1993: vignette 4).

Matilija Dam and Wheeler Gorge, Santa Ynez Mountains, Ventura Co.

Connection: Chapter 1, section on the plate-tectonic evolution of Southern California; chapter 6, section on how river dams block beach sand.

Where: From Ojai, drive north on Highway 33 from its junction with Highway 150. The highway soon enters Wheeler Gorge, a deep canyon cut through the Santa Ynez Mountains by the North Fork of Matilija Creek. After 5.0 miles, turn left onto narrow, winding Matilija Road. Matilija Reservoir will appear to your left. At 0.6 miles from Highway 33, pull over on the left at a small turnout by a narrow, unpaved access road at 34.49194°, −119.31200°. The road leads past the reservoir to the dam about a half-mile away at 34.48507°, −119.30778°. To see more of Wheeler Gorge, return to Highway 33 and continue north for a few miles. If you choose to stay overnight, you'll find several nice campgrounds tucked into this spectacular canyon.

What's There: The rotting concrete of 1948 Matilija Dam and the nearly silt-full reservoir behind it illustrate the blight that present generations are now inheriting from the nation's post–World War II splurge of dam building. From an engineering perspective, this location seems ideal for a dam site because the canyon is so narrow. But the rocks that make up the surrounding mountains are all geologically young, soft sedimentary layers that erode easily. During winter rainstorms, the creeks carry vast amounts of eroded rock debris and thus silt in their reservoirs faster than they would in areas of harder, more erosion-resistant rock. The Santa Ynez Mountains, crushed as they are in the tectonic vise of the Big Squeeze (figure 1.3), are arching upward by as much as a half-inch per year in places. The rivers and creeks have managed to cut downward apace with the rising mountains but have had no time to widen their valleys. They thus lie trapped in deep, narrow canyons—a signature of tectonically recent and rapid uplift. Wheeler Gorge and the surrounding canyons are dramatic examples, as is Malibu Canyon (described several stops above).

Rincon Point on the Ventura–Santa Barbara County Border

Connection: Chapter 5, section on surfing waves at Rincon Point.

Where: From Highway 101, take exit 83 (Bates Road), turn toward the ocean, and park in the lot for Rincon Point at 34.37519°, −119.47633°. A walking trail leads east past private homes to the beach at the east end of the point. Walk west along the beach about one-third of a mile, past the mouth of Rincon Creek, to reach the apex of the point.

What's There: During long-period west swells, the refraction here is so dramatic that the publisher chose a photo of it for the book cover. The west-approaching swells slow down in the shallow water near the point but continue fast in the deep water south of the point, thus bending through about a ninety-degree angle as they wrap around the point. The long right breaks can give surfers rides of a quarter-mile or more. Floods down Rincon Creek have swept sand and rocks into the sea to make the bulging point. If the tide is low, you'll see that the seabed is littered with rocks swept into the sea during floods.

Carpinteria State Beach, Carpinteria, Santa Barbara Co.

Connection: Chapter 3, section on oil formation in Southern California.

Where: Drive to the entrance for Carpinteria State Beach at the south end of Palm Avenue in Carpinteria (34.39251°, −119.52131°) and proceed east past several camping areas and parking lots to the parking area for the tar seeps at 34.38893°, −119.51681°.

What's There: The half-mile of beach from here east to the pier exhibits numerous tar seeps. Tar is just particularly thick crude oil, so it flows more like honey than like motor oil. The tar oozes from the bluffs like black lava and puddles up in wrinkled patterns. Active seeps are shiny; inactive seeps are matte black. The tar originates in the Monterey Formation, an organic-rich package of sedimentary layers, Miocene in age, common throughout Southern California. (See the description for Gaviota State Beach below.) Once oil forms underground, it rises buoyantly upward. If faults or rock layers underground don't block and trap the oil, it can flow all the way to the surface and ooze out as natural seeps. For more about oil and oil production in this region, see Norris (2003: chapter 6).

Santa Barbara Harbor, Santa Barbara Co.

Connection: Chapter 6, section on how human shore structures interfere with longshore drift.

Where: Park along West Cabrillo Boulevard west of Sterns Wharf in downtown Santa Barbara near 34.40994°, −119.69184°, and walk south across the beach.

What's There: You may see a dredge boat working across the harbor near the sand spit at the end of the breakwater. When dredging is active, the boat picks up a slurry of sand and water from the sand spit and pumps it through a two-foot-diameter pipe that runs across the bottom of the harbor to this beach. The pipe then turns and goes east under the beach over a mile beyond Sterns Wharf. The sand is eventually dumped into the surf zone far enough east where the waves, unaffected by the breakwater, can pick it up and keep it moving east. Since its construction in 1930, the breakwater has blocked the natural eastward longshore drift of sand. Within a few years after the breakwater went up, the beaches to the east had become badly eroded while the beaches to the west had grown substantially. As the east-drifting sand works its way around the end of the breakwater, it settles in the quiet water of the harbor mouth, threatening to plug the harbor instead of moving onward to the eastern beaches as it did before. Keeping the harbor open requires dredging 300,000 cubic yards of sand per year. For more, see Norris (2003: chapter 2).

Gaviota State Beach, Santa Barbara Co.

Connection: Chapter 3, section on oil formation in Southern California; chapter 4, section on Miocene rotation of the Western Transverse Ranges.

Where: Take Highway 101 about twenty-four miles west of Goleta and turn off at the entrance for Gaviota State Beach, which comes up just after the highway turns north away from the coast to cross the Santa Ynez Mountains through Gaviota Gorge. Park in the day-use lot (34.47168°, −120.22770°) and walk south under the railroad bridge toward the beach.

What's There: Notice in the cliffs on both sides the rock layers tilting steeply to the south, somewhat like books tilted over on a shelf. It's hard to find rock layers in their original, flat-lying position anywhere in this part of California because of the ongoing tectonic crushing of the Big Squeeze, which distorts and tilts the layers. These rocks belong to the Monterey Formation and are some five million to nineteen million years old, depending on location. As the Continental Borderland formed during the past twenty million years (see chapter 4), numerous deep basins stretched open on the seabed. Plankton, particularly diatoms—plant-like microorganisms with tiny silica shells—rained down abundantly into these basins to create the Monterey Formation. In its purest state, the Monterey Formation is pure *diatomite*—a powdery, porous, snow-white rock made almost entirely of diatom remains. In areas where the Monterey Formation was buried deeply enough, heat and pressure transformed the diatom-rich organic matter within into crude oil, thus forming the source rock for most of Southern California's petroleum-extraction industry. But the layers here have yet another story to tell. Recall the section in chapter 4 on the great clockwise rotation of that big chunk of crust known as the Western Transverse Ranges Block as it migrated from near San Diego to where it is now (figure 4.8). A major piece of evidence for this astonishing rotation and migration comes from magnetic signals preserved in rock units such as the Monterey Formation. See figure 4.9 and the related text for the full story.

Gaviota Gorge, Santa Barbara Co.

Connection: Chapter 1, section on the plate-tectonic evolution of Southern California.

Where: Highway 101 north from Gaviota State Beach crosses the axis of the Santa Ynez Mountains through Gaviota Gorge (34.48824°, −120.22630°).

What's There: The scenic defile of Gaviota Gorge was cut by Gaviota Creek, the only stream that slices all the way across the western Santa Ynez Mountains.

Like Malibu Creek in the Santa Monica Mountains (see description above), Gaviota Creek apparently flowed in more-or-less its current path to the sea before the Santa Ynez Mountains rose. The Santa Ynez Mountains lie in the tectonic region known as the Big Squeeze, where the Big Bend in the San Andreas fault causes the Pacific and the North American tectonic plates to mash into each other (figure 1.3), resulting in geologically rapid uplift and a high risk of earthquakes (figure 3.2). As the mountains squeezed upward (mostly within the past five million years), the creek cut downward apace with uplift to carve Gaviota Gorge. The steep slopes, narrow canyons, and youthful rock strata tilted akimbo throughout this area testify to geologically recent and rapid uplift.

OFFSHORE SITES

Cortes Bank

Connection: Chapter 5, section on surfing at Cortes Bank.

Where: A one-hundred-mile boat trip from San Diego. Bishop Rock, the barely awash summit of Cortes Bank, is at 32.45036°, −119.12476°.

What's There: Probably the largest waves to be found anywhere in Southern California if you go during a long-period northwest swell. The northwest alignment of the bank causes swells arriving from that direction to refract tightly to the shallows atop the bank, focusing their energy to create immense waves (see figure 5.7 and the related text). Under calm conditions, the bank is a magnificent scuba-diving location, although not for the inexperienced or faint-of-heart.

San Clemente Island, Southwest Side

Connection: Chapter 7, section on marine terraces.

Where: The southwest side of San Clemente Island viewed from boat, low-flying airplane, or low-angle view on Google Earth (32.8550°, −118.4690°).

What's There: The most distinct "staircase" of marine terraces anywhere in Southern California extends for miles across San Clemente's wave-hacked southwestern face. Morning light is best because it casts shadows across the terraces. See figure 7.3 for a photograph of the terraces from the island itself, and figure 7.7 for an explanation of how such sequences of marine terraces form.

Two Harbors, Catalina Island

Connection: Chapter 4, section on the tectonic evolution of the Continental Borderland.

Where: Ferries sail regularly from San Pedro to the hamlet of Two Harbors on the north end of Catalina Island. From the boat dock, walk east along the beach about two hundred yards to where outcrops begin at 33.44116°, −118.49647°. A quarter-mile beyond is a pleasant campground, the main accommodation for overnight visitors. East of there, in the half-mile of wave-cut cliffs between the campground and Fisherman's Cove, you'll find more outcrops of interest, all best explored by kayak (rentable at the boat dock) since foot access from the bluff top is difficult.

What's There: In this small area, you'll find several types of rock that illustrate much of the plate-tectonic evolution of the Continental Borderland. According to current theory, the eastern half of the Continental Borderland represents rock that once lay deep underground, in the now extinct subduction zone of the ancient Farallon Plate. As various types of seabed rocks were carried on the plate down into the subduction zone, they underwent low temperature–high pressure metamorphism (see figure 4.10). Starting about twenty million years ago, the Pacific Plate began to capture several pieces of crust from the edge of North America and drag them off to the northwest. A major portion of this captured crust was the Western Transverse Ranges Block, which traveled from near San Diego to where it is now, rotating clockwise in the process (see figure 4.8). As the Western Transverse Ranges Block slid away from this area, rocks deep beneath it bobbed upward to take its place. As all this was happening, the crust also stretched dramatically (see the areas labeled "stretched zone" in figures 4.7 and 4.8). The stretching triggered volcanic eruptions and opened a series of deep marine basins that became receptacles for organic-rich sediments.

Here's how the rocks near Two Harbors fit into that story. The outcrop along the beach between the boat dock and the campground is part of the Catalina Schist—the deep subduction zone rock that bobbed upward as the region stretched and rocks overhead moved off to the northwest. The Catalina Schist is widespread throughout the Continental Borderland and is made of a diverse assortment of rocks produced by metamorphism in a subduction zone. Here it is mostly blueschist—a product of shale metamorphism (figure 4.10a). Just west of the campground, you'll see lots of gray and brown volcanic lava flows and flow-breccias. These represent eruptions that happened as the region stretched. In the sea cliffs between the campground and Fisherman's Cove, you'll find more lava flows, in some places overlain by bright white layers of the Monterey Formation, made mostly of the microscopic remains of ocean plankton called diatoms and forming a rock called *diatomite*. The diatomite deposits accumulated in deep basins formed within the stretching crust. Where buried deeply enough in certain areas, the organic-rich diatomite sediments transformed into crude oil. For related information about the Monterey Formation, see the descriptions for Carpinteria tar seeps and Gaviota State Beach above. For related information about the Catalina Schist, see the descriptions for Dana Point and Aliso Beach County Park above.

Simonton Cove, San Miguel Island

Connection: Chapter 1, section on the immigrant Mexican pebbles and the plate-tectonic evolution of Southern California.

Where: Access to nearly all of San Miguel Island is highly restricted by the park service. To go anywhere other than the beach at Cuyler Harbor and the nearby campground, you need permission from the park service and a ranger escort.

The only accommodation on the island is one primitive campground reached by a steep hike up from the beach. The campground has no water (bring all your water to the island with you). Simonton Cove lies 2.5 miles from the campground on the northwestern side of the island. The outcrop of interest is located at 34.05470°, –120.38639°.

What's There: This is the location of the immigrant Mexican pebbles that I described in chapter 1 (see figure 1.1). The purple-maroon volcanic pebbles abundant in this outcrop (part of the Cañada Formation of Eocene age) are identical in appearance and chemistry to volcanic lava beds in Sonora, Mexico, and to pebble-rich deposits found near San Diego, including in Black's Canyon, La Jolla (see description above). The surrounding geology indicates that the pebbles were flushed into the ocean as part of a deep-water delta near the mouth of the now long extinct Ballena River. The far-flung distribution of the pebbles reflects how the sidling movement of the Pacific Plate northwest past the North American Plate dismembered the pebbly deposits of this ancient river (see figure 1.4).

Glossary

ACCRETED TERRANE A region of imported rock that has come from elsewhere and has been added to the edge of a continent by subduction. Continents grow at their edges by the accretion of terranes, a process responsible for assembling much of western North America, including California. See *accretionary wedge, subduction.*

ACCRETIONARY WEDGE Rock and sediment scraped off the top of a subducting oceanic plate at an ocean trench and added to the adjacent plate to become an accreted terrane. Much of coastal Washington, Oregon, and California is built of accretionary wedge rock scraped off the ancient Farallon Plate and added onto the western edge of the North American Plate. See *accreted terrane, Farallon Plate subduction, subduction zone.*

A-FRAME Surfing term for a wave that breaks in two directions away from a central peak. Usually caused by a wave bending concave in the direction it is traveling in response to the shape of the seabed so that the wave's energy is focused into a central spot. The waves at many famous big-wave surf sites are A-frames.

ANTICLINE An area where layered sedimentary rocks, instead of lying flat, arch upward in the center (picture bending a paperback book with your hands into an upward fold). Economically important because anticlines commonly trap crude oil and natural gas underground. See *reservoir rock, source rock.*

BANK An area of shallow water surrounded by deep water. The main banks offshore of Southern California are Cortes Bank, Tanner Bank, and Thirty-Mile Bank.

BASALT A common black volcanic rock formed from solidified lava and composed of sand-sized crystals, mostly plagioclase feldspar, pyroxene, and olivine.

BASIN A low place, either on land or on the ocean floor, where sediments may accumulate.

BASIN AND RANGE PROVINCE A region of north–south-oriented, fault-bounded mountains separated by sediment-filled valleys (basins), encompassing all of Nevada and portions of adjacent states along with northernmost Mexico. Began forming about twenty million years ago by east–west stretching of the crust and continues to form today.

BEACH COMPARTMENT An area of coastline where beach sand is delivered to the beach by river runoff and bluff erosion, travels along the beach by longshore drift, and eventually leaves the beach down a submarine canyon.

BEDDING A general term for layers of sedimentary rock, also called *strata,* that are originally laid down flat like blankets on a bed, but which may later be faulted, tilted, or folded by tectonic movements.

BIG BEND A 180-mile stretch of the San Andreas fault where the fault turns westerly between Palm Springs and the southern Carrizo Plain. The bend causes the Pacific Plate to push against the North American Plate throughout this area instead of sliding side-by-side past it. The resulting compressive forces have pushed up the Transverse Ranges and make the region highly prone to earthquakes. See *Big Squeeze, Transverse Ranges.*

BIG SQUEEZE The broad region centered on the Big Bend of the San Andreas fault dominated by compressive forces between the Pacific and North American tectonic plates. See *Big Bend, Transverse Ranges.*

BLUESCHIST A type of metamorphic rock formed by low temperature–high pressure metamorphism of shale in a subduction zone. The blue color comes from the metamorphic mineral glaucophane. See *metamorphic rock, schist.*

BLUFF A cliff or steep slope that rises above a beach. Bluffs line roughly half of Southern California's coastline.

BREAKER A breaking wave.

BREAKWATER A structure, usually made of large rock blocks or concrete, designed to protect a harbor or boat anchorage from ocean waves.

CALIFORNIA CURRENT An immense ocean current that flows south past North America's west coast, carrying cool water from the northern Pacific Ocean toward the equator. The California Current is part of a great loop of ocean water that flows clockwise around the North Pacific basin.

CASCADIA SUBDUCTION ZONE The active subduction zone of northernmost California, Oregon, and Washington, where the oceanic Juan de Fuca Plate is plunging eastward under the North American Plate at the Cascadia Trench to form the Cascade Range of active volcanoes.

CATALINA SCHIST A rock formation common in the eastern half of the Continental Borderland and particularly well exposed on Catalina Island. Composed of a wide range of metamorphic rocks interpreted as having once accumulated in an accretionary wedge above the subducting Farallon Plate and which have subsequently undergone low temperature–high pressure metamorphism. Widespread crustal stretching of the Continental Borderland beginning about twenty million years ago allowed these deep-rooted rocks to pop up to the Earth's surface. See *metamorphic core complex, San Onofre Breccia.*

CONEJO VOLCANICS A formation of lava flows and associated volcanic rocks, erupted from sixteen million to thirteen million years ago, that make up large portions of the Santa Monica Mountains. Tiny magnetic minerals in the Conejo Volcanics indicate that the Western Transverse Ranges Block underwent a clockwise rotation of about ninety degrees as the Pacific Plate captured it and shipped it northwest to its current location during the past fifteen million years. See *Western Transverse Ranges Block.*

CONGLOMERATE A sedimentary rock made of rounded gravel or pebbles in a sandy matrix.

CONTINENTAL BORDERLAND The offshore region of Southern California between Point Conception and northern Mexico dominated by deep basins alternating with shallow banks and islands. Consists mostly of foundered blocks of continental crust that were once attached to the North American Plate but that, during the past twenty million years, have transferred over to the Pacific Plate and undergone widespread stretching.

CONTINENTAL SHELF The undersea edge of the continent, sloping gradually seaward from the coastline to about three hundred to six hundred feet of water and ending at the shelf break, where slopes increase down to abyssal ocean depths.

CRUST The outermost layer of the Earth, consisting of either continental crust (twenty to forty miles thick and mostly of granitic composition) or oceanic crust (up to seven miles thick and mostly of basaltic composition).

DEEP TIME An informal term for the millions and billions of years of geologic time.

DIATOMITE A type of sedimentary rock composed mostly of the silica-rich microscopic remains of diatoms: single-celled, microscopic planktonic algae common in ocean surface waters. See *Monterey Formation, plankton.*

DISPERSION In oceanography, the process of ocean waves separating from one another as they travel, with faster waves moving ahead of slower waves. Dispersion of waves from storms produces swells. See *swells.*

EARTHQUAKE Vibrations within the Earth's crust produced by the sudden release of accumulated strain energy when a fault snaps and the rocks on either side leap to a new position.

EDDY A circular movement of a current, usually in a different direction from the main current.

ESTUARY A tidal wetland or saltwater inlet along a shoreline, often formed in a drowned stream valley, where ocean water and fresh stream water mix. Often called *lagoons* or *marshes* along the Southern California coast. Examples: Mugu Lagoon, Penasquitos Marsh.

FARALLON PLATE A large oceanic plate that once filled much of the Pacific Ocean basin and is now mostly extinct as a result of its near total subduction under North America's western edge during the past 140 million years.

FAULT A planar or gently curved fracture in the crust where the rocks on either side have shifted measurably. The energy released when a fault shifts produces earthquakes.

FAULT SCARP A cliff formed by movement along a fault during an earthquake; represents the exposed surface of a fault that penetrates deep into the ground.

FETCH The distance of water across which wind blows to create waves. The greater the fetch, the larger the waves can grow.

FORE-ARC BASIN The region of a subduction zone that lies between the accretionary wedge and the volcanic arc; typically a low area that accumulates sediments shed off the accretionary wedge and/or the volcanic arc. See *accretionary wedge, volcanic arc.*

FORMATION A rock unit with consistent features that allow it to be identified and mapped across a large area, and usually given a specific name. Examples: Monterey Formation, Catalina Schist, San Onofre Breccia.

HALF-LIFE The time required for half of a given amount of radioactive atoms to transform (decay) into another form. Half-lives allow us to determine the ages of rocks and fossils. See *isotope, radiometric dating.*

ICE AGES An informal term for periods of geologic time when the Earth was cooler than normal and when large areas of the continents were covered by ice sheets.

IGNEOUS ROCK Any rock formed by solidification of molten rock, either underground or on the Earth's surface.

ISOTOPE Varieties of a particular chemical element distinguished by different numbers of neutrons in the atomic nucleus. Some isotopes are radioactive and are thus useful for geologic dating. See *half-life, radiometric dating.*

LAGOON. See *estuary.*

LEFT Surfing term for a wave that breaks toward a surfer's left as he or she rides the wave.

LONGSHORE DRIFT Movement of sand parallel to the shoreline due to waves approaching the shore at an angle instead of straight on; also called *longshore transport.* Net longshore drift along the Southern California coast is to the south because ocean waves approach from the northwest most of the time.

MAGMA Molten rock material formed within the Earth; becomes igneous rock upon cooling and solidification. Magma that erupts onto the surface is called *lava.*

MAGNITUDE For earthquakes, a measure of the energy released during an earthquake.

MANTLE The 1,800-mile-thick region between the Earth's crust and core, forming more than three-quarters of the volume of the Earth.

MARINE TERRACE A wave-cut platform raised above the sea by tectonic uplift. See *wave-cut platform.*

MARSH See *estuary.*

METAMORPHIC CORE COMPLEX A geologic region, usually many square miles in area, characterized by metamorphic rocks that have risen to the Earth's surface from several miles underground because the rocks above them have slid laterally away during widespread stretching of the crust.

METAMORPHIC ROCK A rock formed by the alteration, in a solid state, of a preexisting rock by heat and/or pressure and/or fluid interactions deep underground.

MID-OCEAN RIDGES Broad, continuous ridges on the ocean floor, several hundred miles to more than a thousand miles across, with rift valleys running down the centers. The seafloor spreads from the rift valleys, forming new oceanic crust through the eruption of basalt lava. See *pillow basalt, seafloor spreading.*

MINERAL Any naturally occurring inorganic crystalline substance possessing an orderly crystalline structure and a well-defined chemical composition. Aggregations of minerals form rocks.

MONTEREY FORMATION A rock formation widespread in Southern California, ranging from five to nineteen million years in age, consisting mostly of thin white to tan beds of diatomite and forming an important oil source rock. See *diatomite, source rock.*

NORTH AMERICAN CORDILLERA The belt of mountains that stretches from Alaska to Panama along North America's western edge. Includes all the mountains in the western United States from California to the Rockies.

OCEANIC TRENCH A deep, linear depression on the ocean floor formed by subduction, where an oceanic plate bends down underneath an adjacent plate to plunge into the Earth's interior. See *subduction, subduction zone.*

OOZE Sediment consisting of at least 30 percent skeletal remains of tiny planktonic organisms. Oozes commonly accumulate on the deep seabed far from land.

PALEOMAGNETISM Fossil magnetic orientations preserved in certain rocks, usually lava beds and sedimentary layers. Can be used to deduce changes in the direction of the Earth's magnetic field in the past, and the changing position and rotation of parts of the crust caused by tectonic forces. See *Western Transverse Ranges Block.*

PILLOW BASALT (PILLOW LAVA) Bulbous, pillow-shaped lava formations formed where basalt lava erupts underwater. Most of the ocean floor beneath a veneer of sediments is made of pillow basalt produced by seafloor spreading at mid-ocean ridges. See *seafloor spreading.*

PLANKTON Any organism that floats and drifts in the ocean or in fresh water, often microscopic and tremendously abundant. Planktonic remains form a large component of many seabed sediments.

PLATE TECTONICS The theory, confirmed by abundant evidence, that the Earth's outer rocky shell is broken up into several dozen individual plates, fifty to a hundred miles thick, that move and interact to produce earthquakes, volcanoes, and most of the major geographic features of the planet. See *tectonic plate, tectonics.*

RADIOACTIVITY (RADIOACTIVE DECAY) The process whereby an unstable atomic nucleus undergoes fission or releases particles to transform itself into a new chemical element. By convention, radioactive atoms are called *parents* and the products of their transformation (decay) are called *daughters.* See *half-life, radiometric dating.*

RADIOMETRIC DATING The science of determining the ages of rocks and geologic events by measuring the ratio of radioactive parent atoms to their daughter products. See *half-life, radioactivity.*

REFRACTION For ocean waves, the bending of waves as they slow down in water shallower than their wave base. Refraction causes waves to conform somewhat to the shape of the bottom and, in some instances, to focus their energy into especially large breaking waves. See *wave base.*

RELEASING BEND (IN A FAULT) An area where a bend in a side-by-side moving fault allows the rocks on either side to move apart, causing them to sag down to form a low place. See *restraining bend.*

RESERVOIR ROCK A rock formation that both holds abundant crude oil and/or natural gas and is permeable (meaning fluids can flow through it easily), such that the oil or gas can be extracted. See *source rock.*

RESTRAINING BEND (IN A FAULT) An area where a bend in a side-by-side moving fault forces the rocks on either side to push into each other, forcing the Earth's crust upward to make mountains or islands. See *releasing bend.*

RIGHT Surfing term for a wave that breaks toward a surfer's right as he or she rides the wave.

SAN ONOFRE BRECCIA A rock formation, common throughout Southern California, composed of large, angular blocks eroded from now vanished highlands made of Catalina Schist that rose across large portions of the Continental Borderland during Miocene time. See *Catalina Schist.*

SAND Sedimentary particles between one sixteenth and two millimeters in diameter.

SCHIST A common metamorphic rock characterized by distinctive layering formed from the parallel arrangement of flat mineral grains, particularly micas. See *metamorphic rock.*

SEAFLOOR SPREADING The mechanism whereby new ocean floor is created at mid-ocean ridges as two oceanic plates diverge and magma wells up to fill the gap. See *mid-ocean ridges, pillow basalt.*

SEDIMENTARY ROCK A rock formed from eroded pieces of preexisting rocks that have been transported by wind, water, or ice and then deposited and cemented together. Also, any rock formed from particles of biological skeletons or shells, or chemically precipitated out of water.

SEISMIC WAVES Wave-like movements of the Earth's surface and interior produced by earthquakes.

SEISMOGRAPH A ground-motion instrument used to detect and measure seismic waves.

SHALE A fine-grained sedimentary rock made of thin layers of mud, clay, and silt particles.

SOURCE ROCK A sedimentary rock rich in organic material that may transform into crude oil and natural gas under the right conditions deep underground. See *Monterey Formation, reservoir rock.*

SOUTHERN CALIFORNIA BIGHT The broad concave curve of the coastline between Point Conception and San Diego. "Bight" is a nautical term for a large inward curve in a coast that is larger and less confined than a bay.

STRATA Layers of sedimentary rock, originally laid down flat like blankets on a bed, but which may later be faulted, tilted, or folded by tectonic movements; also called *bedding.*

SUBDUCTION The process whereby a moving oceanic plate bends down beneath an adjacent plate and dives into the mantle at an oceanic trench. Characterized by frequent earthquakes, accretion of terranes, and volcanism in the adjacent volcanic arc. See *accreted terrane, accretionary wedge, volcanic arc.*

SUBDUCTION ZONE A region where subduction is occurring.

SWELL SHADOW A region of smaller waves on the down-wave side of an island or shallow bank caused by the island or bank blocking swells. San Clemente, Santa Catalina, and the Northern Channel Islands all cast swell shadows onto the mainland Southern California coast, depending on swell direction.

SWELL WINDOW A region where ocean swells pass unimpeded through gaps in offshore islands, and thus hit the mainland coast with little loss of energy.

SWELLS Large ocean waves generated by storms at sea. Storm-generated swells are the main waves you see breaking on most days along the coast and are the main waves that surfers ride.

TECTONIC PLATE A large region of the Earth's outer shell of rigid rock, fifty to a hundred miles thick, that moves, usually in a different direction from adjacent plates, so that the boundaries between plates are geologically active areas with lots of earthquakes.

TECTONICS The study of processes and forces that cause movement and deformation of the Earth's crust on a large scale. See *plate tectonics.*

TERRANE A block of the Earth's crust, bounded by faults, whose geologic history is distinct from that of adjacent crustal blocks, often because it has traveled from far away. Continents grow by the accretion of terranes to their edges. See *accreted terrane.*

THRUST FAULT A fault in which the rock above an inclined fault has moved up and over the rock below the fault, usually in response to sideways compression.

TRANSVERSE RANGES Collectively, the Santa Ynez, San Gabriel, and San Bernardino mountains, along with associated smaller ranges; they are aligned generally east–west across Southern California, transverse to the mostly northwest–southeast alignment of other mountains in the state; pushed up by compressive forces in the Big Squeeze along the Big Bend. See *Big Bend, Big Squeeze.*

TSUNAMI A large and often destructive wave caused by sudden displacement of ocean water during an undersea earthquake, landslide, volcanic eruption, or (very rarely) meteorite impact.

VOLCANIC ARC A line of active volcanoes that rises parallel to an oceanic trench where subduction is taking place. Examples include the Andes Mountains and the Cascade Range. See *subduction.*

WAVE BASE The depth of water, equal to half the wavelength of the wave, where the orbital motion of passing ocean waves ceases. Waves moving in water deeper than their wave base do not interact with the seabed, whereas waves that enter water shallower than their wave base slow down, refract, and eventually break. See *refraction, wavelength.*

WAVE HEIGHT The vertical distance between the high point (crest) and low point (trough) of an ocean wave.

WAVE ORBIT The orbital motion of water induced by ocean waves passing through an area, decreasing with depth and ceasing at the wave base. See *wave base, wavelength.*

WAVE PERIOD The time, in seconds, between successive wave crests in a series of ocean swells; increases in direct proportion to the amount of energy that winds impart to the ocean. A very useful number, both because it can be measured remotely by sensor buoys and because it is related to how waves will refract and break as they approach the shore.

WAVE-CUT PLATFORM A flat, step-like notch cut into bedrock by wave erosion; becomes a marine terrace if raised above the sea. See *marine terrace*.

WAVELENGTH The horizontal distance between successive wave crests in a set of ocean swells; increases in direct proportion to the amount of energy that winds impart to the ocean.

WESTERN TRANSVERSE RANGES BLOCK (WTRB) The block of the Earth's crust that makes up the Santa Ynez and Santa Monica mountains, the Santa Barbara Channel, the Northern Channel Islands, and all areas in between. Geologic evidence indicates that the WTRB has undergone a dramatic migration and rotation over the past fifteen million years as the Pacific Plate tore it away from the North American continent and shipped it northwest.

Notes on Sources

Most of the science in *Surf, Sand, and Stone* comes from primary sources in the geologic and oceanographic literature, primarily from technical papers in journals, field guidebooks, and edited volumes. The notes below describe the sources that informed particular topics. I list all sources by author and year of publication, corresponding to the bibliography. For any source not listed in the bibliography, including Internet sites, I provide a full citation below.

CHAPTER 1: TIME, FAULTS, AND MOVING PLATES

My explanation of the plate-tectonic evolution of Southern California is based on Atwater (1970, 1998) and Nicholson et al. (1994). The National Park Service provides a good summary of the geology of the northern Channel Islands at www.nps.gov/chis/naturescience/geologicformations.htm. Evidence of the Eocene Ballena River, the northwestward migration of the northern Channel Islands, and the rotation of the Transverse Ranges block comes from Abbott (1999), Abbott and Smith (1978, 1989), Atwater (1998), Kamerling and Luyendyk (1985), Kies and Abbott (1983), and Steer and Abbott (1984). My explanations of the North American Plate—Pacific Plate boundary, including information about active faults and rates of movement, come from Antonelis et al. (1999), Atwater and Stock (1998), Bennett et al. (1999), Kent et al. (2005), Oskin et al. (2001), and Unruh et al. (2003). For maps portraying rates of movement and deformation of western North America based on GPS measurements, see Flesch et al. (2000) and Kreemer et al. (2012). The identification of the Sierran Plate is based on Moores et al. (2006) and Unruh et al. (2003). My speculations about the evolving geography of the southwestern United States (the basis of figure 1.6) are based, in part, on speculations by Dickinson (2009). My information about the Southern California Bight and Continental

Borderland comes from papers in Lee and Normark (2009), along with online materials developed by Bruce Perry, California State University, Long Beach: www.csulb.edu/depts/geology/facultypages/bperry/bperry.html.

CHAPTER 2: TSUNAMIS

My information about the seaports of Los Angeles and Long Beach comes from the port websites. My discussion of the Japanese tsunami of March 11, 2011, is based on the U.S. Geological Survey's summary of the event and on a PBS *NOVA* television program titled "Japan's Killer Quake." My fictional account of the tsunami destruction of the Los Angeles and Long Beach seaports is based on my explorations of the seaport complex using tsunami inundation maps published by the California Emergency Management Agency in 2009, using as a guide damage sustained by the Japanese port of Sendai (Tomita et al. 2011), which experienced 15 to 26 feet of run-up during the March 2011 tsunami despite having a breakwater system comparable to the LA–Long Beach complex. There is a sizable literature on potential tsunamis in the Continental Borderland. My discussion draws on Barberopoulou et al. (2011), Bohannon and Gardner (2004), Borrero et al. (2004), Fisher et al. (2009a, 2009b), Lee et al. (2009), Legg et al. (2002), Locat et al. (2004), Moffatt and Nichol (2007), Normark et al. (2004), and Ryan et al. (2009). Information about historical tsunamis in the Southern California Bight comes from Agnew (1979) and Moffatt and Nichol (2007). Information on tsunamis in the Pacific and Indian oceans comes from the NOAA Center for Tsunami Research (http://nctr.pmel.noaa.gov/index.html) and the NOAA National Geophysical Data Center for Tsunami Hazards (www.ngdc.noaa.gov/hazard/tsu.shtml). My main reference for the Goleta and Palos Verdes slides was Lee et al. (2009).

CHAPTER 3: EARTHQUAKES

The information that begins this chapter on California's oil and the Monterey Formation comes from Harden (2004: 454–461), Sharp and Glasner (1993: chapter 4), and University of Southern California (2013). The scientific literature on Southern California earthquakes is vast, and any one-chapter treatment will leave many things out. My discussion of the earthquake potential of the southernmost San Andreas fault is based on Jones et al. (2008) and on recent lectures at the Scripps Institution of Oceanography by seismologists Neil Driscoll, Kevin Brown, and Yuri Fialko. My discussion of earthquake risk across Southern California comes from the map of Branum et al. (2008). Supercomputer-generated visualizations of major earthquakes along the southern San Andreas fault can be found at the San Diego Supercomputer Center website (http://visservices.sdsc.edu/projects/scec/terashake/). My information on estimated fault slip rates and magnitudes, the basis for table 3.1, comes from Cao et al. (2003: appendices A and B) and from Ryan et al. (2012) for the offshore San Diego Trough fault. Information on historical earthquakes in Southern California comes from the Southern California Earthquake Center (www.data.scec.org/significant/index.html) and from Fisher et al. (2009a: figure 2). My information on earthquakes in the Continental Borderland comes from Astiz and Shearer (2000), Fisher et al. (2009b), and Ryan et al. (2009). My information on the Rose Canyon fault is from Lindvall and Rockwell (1995) and Rockwell (2010). For books about California earthquakes and faults aimed at general readers, see Hough (2004) and Yeats (2001). For a comprehensive field guide to the San Andreas fault, see Lynch (2006).

CHAPTER 4: DISASSEMBLING SOUTHERN CALIFORNIA

Parts of this chapter are adopted, with modification, from my book *Rough-Hewn Land: A Geologic Journey from California to the Rocky Mountains* (University of California Press).

The geologic maps of Thomas Dibblee, available online through the Dibblee Geological Foundation, are essential references for anyone exploring the bedrock geology of Southern California. Norris (2003) is a good resource for anyone interested in the geology of Santa Barbara County and the northern Channel Islands. The story of Southern California's plate-tectonic development comes mostly from Atwater (1998). Tanya Atwater has produced many excellent animations illustrating tectonic developments in Southern California, available online at http://emvc.geol.ucsb.edu/. Additional sources that informed my discussion of tectonics include Fisher et al. (2009a) and Nicholson et al. (1994). For the paleomagnetic evidence that demonstrates tectonic rotation of the Western Transverse Ranges Block, see Champion and Howell (1986), Hornafius (1985), and Kamerling and Luyendyk (1985). For a good summary of all lines of evidence supporting the tectonic rotation of the Western Transverse Ranges Block, see Fritsche et al. (2001). The concept of the California Continental Borderland developing by large-scale extension of the crust (i.e., the metamorphic core complex model) comes from the classic paper of Crouch and Suppe (1993). My information about Santa Catalina Island and the metamorphic rocks there comes from Rowland (1984), supplemented with materials developed by Gary Jacobson of Grossmont College, available online at www.grossmont.edu/people/gary-jacobson/geology-164.aspx. The story of the San Onofre Breccia is summed up by Sharp and Glasner (1993: chapter 4) and by Abbott (1999: 160–165).

CHAPTER 5: WAVES AND SURFING

My information on the discovery of large waves and subsequent surfing expeditions at Cortes Bank is based on Chris Dixon's (2011) book *Ghost Wave.* All quotations are from Dixon (2011) or a Surfline.com press release published on January 8, 2008 (www.surfline.com/surf-news/_13085/). My discussion of wave physics and wave behavior, including the central role of wave period in forecasting wave behavior, comes from Willard Bascom's classic *Waves and Beaches,* a key reference for anyone interested in wave science (Bascom 1980). My explanations of wave refraction and the logic behind the refraction diagrams in figures 5.7, 5.8, and 5.11 are based on classic work that traces back to planning of amphibious military landings during World War II; see Johnson et al. (1948) and Munk and Traylor (1947). Most of what I wrote about surfing waves at the various mainland sites covered in this chapter comes from my investigations of the bathymetry and wave climate of these sites, along with wave-mechanics features produced in 2010 and 2011 by the late surf forecaster Sean Collins, posted on Surfline.com (www.surfline.com/surf-news/mechanics/). See also Collins's Surfline video about Cortes bank (www.surfline.com/surflinetv/how-it-works/how-it-works-cortes-bank_22567). For analysis of wave patterns in the Southern California Bight, see Xu and Noble (2009).

CHAPTER 6: BEACHES AND COASTAL BLUFFS

My information on Point Fermin is based on my explorations of the area and on McNulty (2012). Information about sea-level changes and predicted sea-level rise comes from the National Research Council, Committee on Sea Level Rise in California, Oregon, and Washington (2012). Most of my information about Southern California's beaches, including replenishment history and the amounts of sand supplied by rivers and bluff erosion, comes from Griggs (2010), Griggs et al. (2005), and Patsch and Griggs (2006), all of which I recommend to anyone interested in exploring coastal issues in depth. Additional information about coastal erosion came from California Department of Boating and Waterways (2002), from Hanak and Moreno (2008), and from Staff Report to the California State Lands Commission (2001). Until recently, most geologists assumed that rivers supplied most of Southern California's beach sand. But in 2005, two independent research groups at the University of California San Diego concluded that erosion of coastal bluffs may supply at least half of the sand on San Diego's beaches (see Young and Ashford 2006, Haas and Driscoll 2005). My information about construction permits in the coastal zone comes from the California Coastal Commission (CCC) website (www.coastal.ca.gov/), supplemented by phone interviews with Eric Stevens of the San Diego office of the CCC. My information about seawall construction comes from an interview with Walter Crampton of TerraCosta Consulting Group. My sections about Matilija Dam simplify the complex and ongoing story behind the dam and efforts to remove it. Interested readers should consult the website of the Matilija Dam Ecosystem Restoration Project (www.matilijadam.org/), which contains extensive references, including the 2000 Bureau of Reclamation report on the feasibility of dam removal. My information on San Diego's 2012 beach replenishment project comes from the San Diego Regional Beach Sand Project website (www.sandag.org/index.asp?projectid=358&fuseaction=projects.detail), and the 2011 environmental impact report for the project (AECOM 2011). See Crampton (2005) for an argument about the necessity of committing to long-term sustained beach replenishment in Southern California. Readers interested in how California communities can adapt to sea-level rise should consult Russell and Griggs (2012).

CHAPTER 7: SEA-LEVEL CHANGES AND THE ICE AGES

My information on Southern California marine terraces is based largely on papers by Dan Muhs (Muhs et al. 2002, 2004, 2006, 2012), along with Abbott

(1999: chapter 9), Harden (2004: chapter 15), and Norris (2003). My discussion of dating marine terraces is based, in part, on Muhs et al. (2004). My information on San Diego marine terraces is from the geologic maps of Kennedy and Tan (2005a, 2005b), along with Kern and Rockwell (1992). For the story of the ice ages from Louis Agassiz through James Croll, Milutin Milankovitch, and the Deep Sea Drilling Project, I relied primarily on Macdougall (2004, an outstanding work that I recommend to anyone interested in learning more about the ice ages), along with Shubin (2013: chapter 9) and Williams et al. (2010). Hays et al. (1976) is the classic paper that confirmed the connection between Milankovitch cycles and the ice ages. My information about the magnitude and timing of sea-level changes during the ice ages is from Kennedy and Tan (2005a) and Peltier and Fairbanks (2006). My information on coastal wetlands comes from Deen and Von (2008) and from the websites of the California Resources Agency (http://resources.ca.gov/), California Wetland Information System (http://ceres.ca.gov/wetlands/), and the San Elijo Lagoon Conservancy (www.sanelijo.org).

AFTERWORD

My information on increasing wave heights offshore of Southern California comes from Seymour (2011), as discussed in Russell and Griggs (2012). See also Ruggiero et al. (2009) for evidence of increasing wave heights off the Pacific coast in recent years. For information on the effects of sea rise on the California coast, see Hanak and Moreno (2008) and Russell and Griggs (2012). My statements about decreasing future precipitation across the American Southwest are based on analyses by the Environmental Protection Agency (www.epa.gov/climatechange/impacts-adaptation/southwest.html). My information about Israeli agriculture comes from the Israel Export and International Cooperation Institute (www.moag.gov.il/agri/files/Israel's_Agriculture_Booklet.pdf). The section on desalination comes from the following: a 2014 news article by Alisa Odenheimer and James Nash, "Israeli Desalination Shows California Not to Fear Drought" (*Bloomberg News*); a 2014 news article by Paul Rogers, "Nation's Largest Ocean Desalination Plant" (*San Jose Mercury News*); and an environmental impact report for the Carlsbad desalination plant by Graham (2005).

Bibliography

Abbott, P.L., 1999. The Rise and Fall of San Diego: 150 Million Years of History Recorded in Sedimentary Rocks. San Diego, CA: Sunbelt.

Abbott, P.L., and T.E. Smith, 1978. Trace-element comparison of clasts in Eocene conglomerates, southwestern California and northwestern Mexico. Journal of Geology, 86, 753–762.

Abbott, P.L., and T.E. Smith, 1989. Sonora, Mexico, source for the Eocene Poway Conglomerate of Southern California. Geology, 17, 329–332.

AECOM, 2011. Environmental Assessment/Final Impact Report, San Diego Regional Beach Sand Project II, prepared for the San Diego Association of Governments. www.sandag.org/uploads/projectid/projectid_358_14427.pdf

Agassiz, L., 1840. Studies on Glaciers. Preceded by the Discourse of Neuchâtel. [Translated from the French and edited by Albert V. Carozzi. New York, NY: Hafner, 1967.]

Agnew, D., 1979. Tsunami history of San Diego. In: Abbott, P., and W. Elliott, editors, Earthquakes and Other Perils: San Diego Region. San Diego, CA: San Diego Association of Geologists. pp. 117–122.

Antonelis, K., D.J. Johnson, M.M. Miller, and R. Palmer, 1999. GPS determination of current Pacific–North American Plate motion. Geology, 27, 299–302.

Astiz, L., and P.M. Shearer, 2000. Earthquake locations in the Inner Continental Borderland, offshore Southern California. Bulletin of the Seismological Society of America, 90, 425–449.

Atwater, T., 1970. Implications of plate tectonics for the Cenozoic tectonic evolution of western North America. Geological Society of America Bulletin, 81, 3513–3536.

Atwater, T., 1998. Plate tectonic history of Southern California with emphasis on the western Transverse Ranges and Santa Rosa Island. In: Weigand, P.W., editor, Contributions to the Geology of the Northern Channel Islands,

Southern California. American Association of Petroleum Geologists Pacific Section, MP 45. pp. 1–8.

Atwater, T., 2003. When the Plate Tectonic Revolution Met Western North America. In: Oreskes, N., editor, Plate Tectonics: An Insider's History of the Modern Theory of the Earth. Cambridge, MA: Westview Press. pp. 243–263.

Atwater, T., and J.M. Stock, 1998. Pacific–North America plate tectonics of the Neogene southwestern United States: An update. International Geology Review, 40, 373–402.

Barberopoulou, A., M.R. Legg, B. Uslu, and C.E. Synolakis, 2011. Reassessing the tsunami risk in major ports and harbors of California I: San Diego. Natural Hazards, 58, 479–496.

Barnard, P.L., D.M. Hanes, R.G. Kvitek, and P.J. Iampietro, 2006. Sand Waves at the Mouth of San Francisco Bay, California. U.S. Geological Survey Scientific Investigations Map 2944.

Bascom, W., 1980. Waves and Beaches: The Dynamics of the Ocean Surface. New York, NY: Anchor.

Bennett, R.A., J.L. Davis, and B.P. Wernicke, 1999. Present-day pattern of Cordilleran deformation in the Western United States. Geology, 27, 371–374.

Bohannon, R.G., and J.V. Gardner, 2004. Submarine landslides of San Pedro Escarpment, southwest of Long Beach, California. Marine Geology, 203, 261–268.

Borrero, J.C., M.R. Legg, and C.E. Synolakis, 2004. Tsunami sources in the Southern California Bight. *Geophysical Research Letters*, 31, L13211.

Branum, D., S. Harmsen, E. Kalkan, M. Petersen, and C. Wills, 2008. Earthquake Shaking Potential for California. California Geological Survey Map Sheet 48 (revised).

Bryson, B., 2004. A Short History of Nearly Everything. New York, NY: Broadway Books.

California Department of Boating and Waterways, 2002. California Beach Restoration Study. www.dbw.ca.gov/Environmental/BeachReport.aspx

Cao, T., W.A. Bryant, B. Rowshandel, D. Branum, and C. Wills, 2003. The revised 2002 California probabilistic seismic hazard maps. Online publication of the California Geological Survey: www.consrv.ca.gov/cgs/rghm/psha/fault_parameters/pdf/Documents/2002_CA_Hazard_Maps.pdf

Champion, D.E., and D.G. Howell, 1986. Paleomagnetism of Cretaceous and Eocene strata, San Miguel Island, California Borderland, and the northward translation of Baja California. Journal of Geophysical Research: Solid Earth, 91, 11557–11570.

Crampton, W., 2005. A different perspective on the concept of planned retreat. In: Magoon, O.T., H. Converse, B. Baird, B. Jines, and M. Miller-Henson, editors, California and the World Ocean '02: Revisiting and Revising California's Ocean Agenda. Reston, VA: American Society of Civil Engineers. pp. 417–426.

Croll, J., 1864. On the physical cause of the change of climate during geological epochs. Philosophical Magazine 28(187), 121–137.

Crouch, J.K., and J. Suppe, 1993. Late Cenozoic tectonic evolution of the Los Angeles basin and inner California borderland: A model for core complex-like crustal extension. Geological Society of America Bulletin, 105, 1415–1434.

Deen, P., and L. Von, 2008. Coastal lagoons in northern San Diego County: Recipes for multi-use and restoration. In: Trujillo, A.P., editor, Geoscience Investigations in Northern San Diego County and Beyond: Student-Directed Explorations. Field Trip Guidebook: National Association of Geoscience Teachers Far Western Section Conference, March 14–16, 2008, Palomar College.

Dickinson, W.R., 2003. The Coming of Plate Tectonics to the Pacific Rim. In: Oreskes, N., editor, Plate Tectonics: An Insider's History of the Modern Theory of the Earth. Cambridge, MA: Westview Press. pp. 264–287.

Dickinson, W.R., 2009. Anatomy and global context of the North American Cordillera. In: Mahlburg Kay, S., V. A. Ramos, and W.R. Dickinson, editors, Backbone of the Americas: Shallow Subduction, Plateau Uplift, and Ridge and Terrane Collision. GSA Memoirs 204. Boulder, CO: Geological Society of America. pp. 1–29.

Dixon, C., 2011. Ghost Wave: The Discovery of Cortes Bank and the Biggest Wave on Earth. San Francisco, CA: Chronicle Books.

Fisher, M.A., V.E. Langenheim, C. Nicholson, H.F. Ryan, and R.W. Sliter, 2009a. Recent developments in understanding the tectonic evolution of the Southern California offshore area: Implications for earthquake-hazard analysis. In: Lee, H.J., and W.R. Normark, editors, Earth Science in the Urban Ocean: The Southern California Continental Borderland. GSA Special Papers 454. Boulder, CO: Geological Society of America. pp. 229–250.

Fisher, M.A., C.C. Sorlien, and R.W. Sliter, 2009b. Potential earthquake faults offshore Southern California, from the eastern Santa Barbara Channel south to Dana Point. In: Lee, H.J., and W.R. Normark, editors, Earth Science in the Urban Ocean: The Southern California Continental Borderland. GSA Special Papers 454. Boulder, CO: Geological Society of America. pp. 271–290.

Flesch, L.M., W.E. Holt, A.J. Haines, and B. Shen-Tu, 2000. Dynamics of the Pacific–North American Plate Boundary in the western United States. Science, 287, 834–836.

Fritsche, A.E., P.W. Weigand, I.P. Colburn, and R.L. Harma, 2001. Transverse/peninsular ranges connections: Evidence for the incredible miocene rotation. In: Dunne, G., and J. Cooper, editors, Geologic Excursions in Southwestern California. Fullerton, CA: Pacific Section, Society for Sedimentary Geology. pp. 101–145.

Graham, J.B., 2005. Marine biological considerations related to the reverse osmosis desalination project at the Encina Power Plant, Carlsbad, CA. Report submitted to Poseidon Resources, April 4, 2005. http://carlsbaddesal.com/Websites/carlsbaddesal/images/eir/Graham.pdf

Griggs, G., 2010. Introduction to California's Beaches and Coast. Berkeley, CA: University of California Press.

Griggs, G.B., K.B. Patsch, and L.E. Savoy, 2005. Living with the Changing California Coast. Berkeley, CA: University of California Press.

Haas, J., and N. Driscoll, 2005. Sources of Beach Sand in the Oceanside Littoral Cell. www.coastalconference.org/h20/2005_presentations.php

Hanak, E., and G. Moreno, 2008. California Coastal Management with a Changing Climate. Public Policy Institute of California. www.ppic.org/content/pubs/report/R_1108GMR.pdf

Harden, D.R., 2004. California Geology, 2nd ed. Saddle River, NJ: Pearson–Prentice Hall.

Hays, J.D., J. Imbrie, and N.J. Shackleton, 1976. Variations in the Earth's Orbit: Pacemaker of the Ice Ages. Science, 194, 1121–1132.

Hickey, B.M., 1992. Circulation over the Santa Monica–San Pedro basin and shelf. Progress in Oceanography, 30, 37–115.

Hornafius, J.S., 1985. Neogene tectonic rotation of the Santa Ynez Range, western Transverse Ranges, California, suggested by paleomagnetic investigation of the Monterey Formation. Journal of Geophysical Research: Solid Earth, 90, 12503–12522.

Hough, S.E., 2004. Finding Fault in California: An Earthquake Tourist's Guide. Missoula, MT: Mountain Press.

Johnson, J.W., M.P. O'Brien, and J.D. Isaacs, 1948. Graphical construction of wave refraction diagrams. U.S. Navy Hydrographic Office Publication 605.

Jones, L.M., R. Bernknoph, D. Cox, J. Goltz, K. Hudnut, D. Mileti, S. Perry, D. Ponti, K. Porter, M. Reichle, H. Seligson, K. Shoaf, J. Treiman, and A. Wein, 2008. The Shakeout Scenario. U.S. Geological Survey Open File Report 2008–1150.

Kamerling, M.J., and B.P. Luyendyk, 1985. Paleomagnetism and Neogene tectonics of the northern Channel Islands, California. Journal of Geophysical Research: Solid Earth, 90, 12485–12502.

Kennedy, M.P., and S.S. Tan, 2005a. Geologic Map of the Oceanside 30'x60' Quadrangle, California. Sacramento, CA: California Department of Conservation.

Kennedy, M.P., and S.S. Tan, 2005b. Geologic Map of the San Diego 30'x60' Quadrangle, California. Sacramento, CA: California Department of Conservation.

Kent, G.M., J.M. Babcock, N.W. Driscoll, A.J. Harding, J.A. Dingler, G.G. Seitz, J.V. Gardner, L.A. Mayer, C.R. Goldman, A.C. Heyvaert, R.C. Richards, R. Karlin, C.W. Morgan, P.T. Gayes, and L.A. Owen, 2005. 60 k.y. record of extension across the western boundary of the Basin and Range Province: Estimate of slip rates from offset shoreline terraces and a catastrophic slide beneath Lake Tahoe. Geology, 33, 365–368.

Kern, J.P., and T.K. Rockwell, 1992. Chronology and deformation of Quaternary marine shorelines, San Diego County, California. In: Quaternary Coasts of the United States: Marine and Lacustrine Systems. Society of Economic Paleontologists and Mineralogists Special Publication 48. pp. 377–382.

Kies, R.P., and P.L. Abbott, 1983. Rhyolite clast populations and tectonics in the California Continental Borderland. Journal of Sedimentary Petrology, 53, 461–475.

Kreemer, C., W.C. Hammond, G. Blewitt, A.A. Holland, and R.A. Bennett, 2012. A geodetic strain rate model for the Pacific–North American plate boundary, western United States. Nevada Bureau of Mines and Geology Map 178, scale 1:1,500,000.

Lee, H.J., H.G. Greene, B.D. Edwards, M.A. Fisher, and W.R. Normark, 2009. Submarine landslides of the Southern California Borderland. In: Lee, H.J., and W.R. Normark, editors, Earth Science in the Urban Ocean: The Southern California Continental Borderland. GSA Special Papers 454. Boulder, CO: Geological Society of America. pp. 251–270.

Lee, H.J., and W.R. Normark, editors, 2009. Earth Science in the Urban Ocean: The Southern California Continental Borderland. GSA Special Papers 454. Boulder, CO: Geological Society of America.

Legg, M.R., J.C. Borrero, and C.E. Synolakis, 2002. Evaluation of tsunami risk to Southern California coastal cities. EERI-FEMA NEHRP Professional Fellowship Report. National Earthquake Hazards Reduction Program, funded by the Federal Emergency Management Agency and administered by the Earthquake Engineering Research Institute.

Lindvall, S.C., and T.K. Rockwell, 1995. Holocene activity of the Rose Canyon fault zone in San Diego, California. Journal of Geophysical Research: Solid Earth, 100, 24121–24132.

Locat, J., H.J. Lee, P. Locat, and J. Imran, 2004. Numerical analysis of the mobility of the Palos Verdes debris avalanche, California, and its implication for the generation of tsunamis. Marine Geology, 203, 269–280.

Luyendyk, B.P., M.K. Kamerling, R.R. Terres, and J.S. Hornafius, 1985. Simple shear of Southern California during Neogene time suggested by paleomagnetic declinations. Journal of Geophysical Research: Solid Earth, 90, 12454–12466.

Lynch, D.K., 2006. Field Guide to the San Andreas Fault. Available online through Thule Scientific: www.thulescientific.com/

Macdougall, D., 2004. Frozen Earth: The Once and Future Story of Ice Ages. Berkeley, CA: University of California Press.

McNulty, B., 2012. Geology of the Palos Verdes Peninsula, Los Angeles, CA. Field guide published by California State University Dominguez Hills. www4.csudh.edu/Assets/CSUDH-Sites/Earth/docs/PalosVerdesfieldguide.pdf

McPhee, J., 1989. The Control of Nature. New York, NY: Farrar, Straus and Giroux.

McPhee, J., 1998. Annals of the Former World. New York, NY: Farrar, Straus and Giroux.

Meldahl, K.H., 2011. Rough-Hewn Land: A Geologic Journey from California to the Rocky Mountains. Berkeley, CA: University of California Press.

Milankovitch, M., 1920. A mathematical Theory of Thermal Phenomena Caused by Solar Radiation. [*Theorie mathematique des phenomenes thermiques produits par la radiation solaire.*] Paris, France: Gauthier-Villars.

Milankovitch, M., 1941. Canon of insolation and the ice-age problem. [Originally published in German as *Kanon der Erdbestrahlung und seine Anwendung auf das Eiszeitenproblem.* English translation from Israel Program for Scientific Translations, 1969.]

Moffatt and Nichol [consulting firm], 2007. Tsunami Hazard Assessment for the Ports of Long Beach and Los Angeles: Final Report. M&N File 4839–169.

Moores, E.M., J. Wakabayashi, J. Unruh, and S. Waechter, 2006. A transect spanning 500 million years of active plate margin history: Outline and field trip guide. In: Prentice, C.S., J.G. Scotchmoor, E.M. Moores, and J.P. Kiland, editors, 1906 San Francisco Earthquake. GSA Field Guides 7. Boulder, CO: Geological Society of America. pp. 373–413.

Mount, J., editor, 2010. Geology and Geomorphology of Eastern Santa Cruz Island [online field guide]. www.geology.ucdavis.edu/~shlemonc/trips/SantaCruz_10/fieldguide.htm

Muhs, D.R., K.R. Simmons, G.L. Kennedy, K.R. Ludwig, and L.T. Groves, 2006. A cool eastern Pacific Ocean at the close of the Last Interglacial complex. Quaternary Science Reviews, 25, 235–262.

Muhs, D.R., K.R. Simmons, G.L. Kennedy, and T.K. Rockwell, 2002. The last interglacial period on the Pacific coast of North America: Timing and paleoclimate. Geological Society of America Bulletin, 114, 569–592.

Muhs, D.R., K.R. Simmons, R.R. Schumann, L.T. Groves, J.X. Mitrovica, and D. Laurel, 2012. Sea-level history during the Last Interglacial complex on San Nicolas Island, California: Implications for glacial isostatic adjustment processes, paleozoogeography and tectonics. Quaternary Science Reviews, 37, 1–25.

Muhs, D.R., J.F. Wehmiller, K.R. Simmons, and L.L. York, 2004. Quaternary sea level history of the United States. Developments in Quaternary Science, 1, 147–176.

Munk, W.H., and M.A. Traylor, 1947. Refraction of Ocean Waves: A Process Linking Underwater Topography to Beach Erosion. Journal of Geology, 55, 1–26.

National Research Council, Committee on Sea Level Rise in California, Oregon, and Washington, 2012. Sea-Level Rise for the Coasts of California, Oregon, and Washington: Past, Present, and Future. Washington, DC: National Academies Press.

Nicholson, C., C.C. Sorlien, T. Atwater, J.C. Crowell, and B.P. Luyendyk, 1994. Microplate capture, rotation of the western Transverse Ranges, and initiation of the San Andreas transform as a low-angle fault system. Geology, 22, 491–495.

Normark, W.R., M. McGann, and R. Sliter, 2004. Age of Palos Verdes submarine debris avalanche, Southern California. Marine Geology, 203, 247–259.

Norris, R.M., 2003. The Geology and Landscape of Santa Barbara County, California, and Its Offshore Islands. Santa Barbara Museum of Natural History Monographs 3.

Oskin, M., J.M. Stock, and A. Martín-Barajas, 2001. Rapid localization of Pacific–North America plate motion in the Gulf of California. Geology, 29, 459–462.

Patsch, K., and G. Griggs, 2006. Littoral Cells, Sand Budgets, and Beaches: Understanding California's Shoreline. Published jointly by the University of California at Santa Cruz, the California Department of Boating and Waterways, and the California Coastal Sediment Management Workgroup. www.dbw.ca.gov/csmw/PDF/LittoralDrift.pdf

Peltier, W.R., and R.G. Fairbanks, 2006. Global glacial ice volume and Last Glacial Maximum duration from an extended Barbados sea level record. Quaternary Science Reviews, 25, 3322–3337.

Piotrowski, D., 2007. A tsunami strikes San Pedro and Wilmington, 1868. In: Channel Crossings (Winter 2007). San Pedro, CA: Los Angeles Maritime Museum. pp. 4–5.

Rockwell, T.K., 2010. The Rose Canyon Fault Zone in San Diego. Fifth International Conference on Recent Advances in Geotechnical Earthquake Engineering and Soil Dynamics. May 24–29, San Diego, CA. Paper 7.06c.

Rowland, S.M., 1984. Geology of Santa Catalina Island. California Geology, 37, 239–251.

Ruggiero, P., P.D. Komar, and J.C. Allen, 2009. Increasing wave heights and extreme value projections: The wave climate of the U.S. Pacific Northwest. Coastal Engineering, 57, 539–552.

Russell, N., and G. Griggs, 2012. Adapting to sea level rise: A guide for California's coastal communities. Public Interest Environmental Research Program of the California Energy Commission. http://calost.org/pdf/announcements/Adapting%20to%20Sea%20Level%20Rise_N%20Russell_G%20Griggs_2012.pdf

Ryan, H.F., J.E. Conrad, C.K. Paull, and M. McGann, 2012. Slip rate on the San Diego Trough Fault Zone, inner California Borderland, and the 1986 Oceanside earthquake swarm revisited. Bulletin of the Seismological Society of America, 102, 2300–2312.

Ryan, H.F., M.R. Legg, J.E. Conrad, and R.W. Sliter, 2009. Recent faulting in the Gulf of Santa Catalina: San Diego to Dana Point. In: Lee, H.J., and W.R. Normark, editors, Earth Science in the Urban Ocean: The Southern California Continental Borderland. GSA Special Papers 454. Boulder, CO: Geological Society of America. pp. 291–315.

Seymour, R.J., 2011. Evidence for changes to the northeast Pacific wave climate. Journal of Coastal Research, 27, 194–201.

Sharp, R.P., and A.F. Glasner, 1993. Geology Underfoot in Southern California. Missoula, MT: Mountain Press.

Shubin, N., 2013. The Universe Within: The Deep History of the Human Body. New York, NY: Pantheon.

Staff Report to the California State Lands Commission, 2001. Shoreline Protective Structures Report. www.slc.ca.gov/Reports/Shoreline_Home_Page.html

Steer, B.L., and P.L. Abbott, 1984. Paleohydrology of the Eocene Ballena Gravels, San Diego County, California. Sedimentary Geology, 38, 181–216.

Tomita, T., G.-S. Yeon, D. Tatsumi, O. Okamoto, and H. Kawai, 2011. Port damage from tsunami of the Great East Japan Earthquake. In: Tamura, K., editor, Proceedings of the 43rd Joint Meeting of U.S.–Japan Panel on Wind and Seismic Effects, August 29–30. Technical Note of PWRI 4217. Tsukuba, Japan: Public Works Research Institute. pp. 179–183.

University of Southern California, 2013. Powering California: The Monterey Shale & California's Economic Future. Los Angeles, CA: USC Price School of Public Policy and Global Energy Network. http://communicationsinstitute.org/Monterey_Shale_Report_Final_130328.pdf

Unruh, J., J. Humphrey, and A. Barron, 2003. Transtensional model for the Sierra Nevada Frontal Fault System, eastern California. Geology, 33, 327–330.

Williams, B., M.P. Clough, M. Stanley, and C. Cervato, 2010. Ice ages: An alien idea. The Story Behind the Science Project. Washington, DC: National Science Foundation. www.storybehindthescience.org/pdf/iceages.pdf

Xu, J.P., and M.A. Noble, 2009. Variability of the Southern California wave climate and implications for sediment transport. In: Lee, H.J., and W.R. Normark, editors, Earth Science in the Urban Ocean: The Southern California Continental Borderland. GSA Special Papers 454. Boulder, CO: Geological Society of America. pp. 171–191.

Yeats, R.S., 2001. Living with Earthquakes in California: A Survivor's Guide. Corvallis, OR: Oregon State University Press.

Young, A.P., and S.A. Ashford, 2006. Application of airborne LIDAR for seacliff volumetric change and beach-sediment budget contributions. Journal of Coastal Research, 22, 307–318.

Index

A-frame wave, 111
accreted terrancs, 66–69
accretionary wedge, 65(fig.), 70, 71, 72(fig.). 73, 78(fig.). *See also* subduction
Agassiz, Louis, 154–157
age of the Earth, 2–3
Aliso Beach County Park, 183
amino acid racemization, 148–149
amphibole schist, 84(fig.), 85
amphibolite, 84(fig.), 85
anticlines, 43, 45, 187–188
assembly of California by plate movements, 2, 61–69

Baja California, 5(fig.), 6(fig.), 8(fig.), 10(fig.), 12, 15(fig.), 18, 74(fig.), 94
Baker, Grant, 102, 104
Ballena River, 4(fig.), 7, 8(fig.), 181, 193
basalt. *See* pillow basalt, ophiolites
Basin and Range Province, 6(fig.), 10–12, 18, 49(fig.), 52(fig.), 73, 74(fig.), 82–83
  geologic formation of, 74(fig.)
  metamorphic core complexes in, 82–83
bathymetry, 37, 38(fig.), 97
beach compartments, 135–137, 138(fig.)
beach replenishment, 126, 137, 140–142, 173–174
beaches, 129–142
  artificial replenishment of, 137, 140–142, 173–174
  effect of breakwaters on, 133–135
  effect of dams on, 130–131, 132(fig.)
  effect of seawalls on, 127, 131–133
  effect of submarine canyons on, 135–137
  longshore drift along, 133–135
  revenue from, 129
  seasonal changes in size of, 129–130
  shrinkage of, 130–142
  sources of sand for, 126(fig.), 130–135
bentonite, 120–121
Big Bend, 6(fig.), 9, 10(fig.), 11, 35, 43–45, 48(fig.), 53–55, 131, 186–187, 188, 190–191
Big Squeeze. *See* Big Bend
Bishop Rock (Cortes Bank), 89, 90, 102, 103(fig.), 104
Black's Beach, 91(fig.), 106–108, 124(fig.), 176, 180–181
Black's Canyon, 181
blueschist, 84(fig.), 85, 86(fig.), 87(fig.)
bluff erosion, 121–129, 131–133, 173, 181
breccia, 86
Brown, Robert, 102
Bureau of Reclamation, 139–140

California Coastal Commission, 119, 127–129, 206
California Current, 18–21
carbon-14 dating, 148

Carpinteria oil seeps, 44, 189
Cascade Range, 64(fig.), 65(fig.), 70–71
Cascadia Subduction Zone (Cascadia Trench), 35, 64(fig.), 65(fig.), 69–71, 76, 83
Catalina fault, 28, 34(fig.), 35, 88
Catalina Island, 28, 34(fig.), 35, 80–88, 191–192
Catalina Schist, 86(fig.), 182–183, 192
Channel Islands. *See* Northern Channel Islands
Channel Islands Harbor, 134(fig.), 135, 187
Channel Islands National Park, 188
climate change, 174–176
Coast Range (Oregon), 69–70
coastal wetlands, 165–170
Cocos Plate, 74(fig.), 76
Conejo Volcanics, 80–81
conglomerate, 3
Continental Borderland, 6(fig.), 15(fig.), 16–23, 25–41, 44, 77, 80–88, 102, 106
  geologic development of, 76–88
  metamorphic rocks of, 83–86
  stretching of inner (eastern) portion, 80–88
  tsunami risk in, 25–41
Cortes Bank, 17, 89–90, 91(fig), 101–105, 191
Croll, James, 156(fig.), 157–161
crust, continental versus oceanic, 16–17
Crystal Cove State Park, 183
cyanobacteria, 164n

dams, 130–131, 132(fig.), 137–140
Dana Point, 182–183
Darwin, Charles, 155
De Charpentier, Jean, 155
Deep Sea Drilling Project, 161–164
deep time. *See* geologic time
desalination of seawater, 175–176
diatomite, 44, 190, 192
diatoms, 44, 190, 192
Dickinson, William, 71
dispersion. *See* wave dispersion

Earth, age of, 2–3
earthquakes
  computer models of, 51–54
  1857 Fort Tejon, 47, 51, 52(fig.)
  evaluating risk from, 47–48
  frequency of, 45
  largest in California history, 51–52
  magnitude, 50–51 (box)
  1906 San Francisco, 51, 52(fig.)
  preparation for, 172
  recurrence interval, 55–57
  relation to tectonic plates, 45–46
  risk from. *See* earthquake risk
  role in assembling California, 69
  role in creating oil traps, 44–45
  role in creating topography, 14, 43–44
  along San Andreas fault, 47–53
  in Southern California, 14–15, 47–53
earthquake risk
  in Big Bend/Big Squeeze, 48(fig.), 53–55
  in coastal counties, 53–59
  in sediment filled basins, 53, 54(fig.)
  across Southern California, 47–53
  along southernmost San Andreas fault, 47–53
Eastern California Shear Zone, 10(fig.), 12, 52(fig.)
eclogite, 84(fig.), 85
El Niño-Southern Oscillation (ENSO), 141
erratic boulders, 154
erosion of coastal bluffs, 124(fig.), 126(fig.), 131–133, 173

Farallon Plate, 8(fig.), 73–77, 81–88, 192
  and formation of the Basin and Range, 73–76
  and formation of the San Andreas fault, 73–76
  metamorphism of during subduction, 81–88, 192
Farallon-Pacific Ridge, 74(fig.), 75, 76
faults
  bends in, 33–35. *See also:* releasing bends, restraining bends
  and earthquakes. *See* earthquakes
  San Andreas. *See* San Andreas fault
  slip rates of, 55–59
  in Southern California(map), viii–ix
  along Southern California coast, 56(table)
fetch, 91, 95(fig.), 100
foraminifera, 161–164
fore-arc basin, 65(fig.), 72(fig.), 73
Franciscan Complex, 71, 72(fig.), 73
future geography of western North America, 9–14

Gaviota Gorge, 190
Gaviota State Beach, 190
geologic dating, 146–148
  amino acid racemization, 148–149
  carbon-14, 148

isotopes used in, 147
radioactive decay, 146–148
uranium-thorium, 148
geologic time, 1–7, 171
geologic time scale, xii
Gerlach, Brad, 102, 104
Goleta Beach (Santa Barbara), 123
Goleta Slide, 36–40
Global Positioning System (GPS) used to measure earth movements, 5, 9–12, 47n
Glomar Challenger (ship), 163(box)
granite, 17(fig.), 48, 51(box), 71, 72(fig.), 73, 122
Great Valley Group, 72(fig.), 73, 78(fig.)
greenschist, 84(fig.), 85, 87(fig.)
greenstone 84(fig.), 85
Gulf of California, 5, 7, 10(fig.), 12, 15(fig.), 74(fig.), 76

hard stabilization, 126–129
hold down (surfing), 104

ice ages, 102, 110, 121, 136(box), 143–170
causes, 157–164
dating of, 158, 159, 161–164
discovery, 153–157
evidence from Deep Sea Drilling Project, 161–164
and formation of coastal wetlands, 165–170
frequency in Earth history, 165(box)
in future, 164–165
Milankovitch cycles, 161–165
oxygen isotopes, 150(fig.), 161–164
and sea level changes, 149–154
isostatic equilibrium, 16
isotopes, 147, 161–164

Japan Trench, 26
Juan de Fuca Plate, 10(fig.), 64(fig.), 65(fig.), 69–71, 73, 74(fig.), 76, 79(fig.)

lagoons. *See* wetlands
La Jolla Canyon, 58(fig.), 106, 107(fig.), 111, 135, 138(fig.), 141, 180
Last Glacial Maximum, 122(fig.), 150(fig.), 154, 158, 167
lithosphere, 9n, 65(fig.)
Lituya Bay (Alaska), 31, 32(fig.), 37, 41
Long, Greg, 102, 105
longshore current, 19
longshore drift, 133–135, 138(fig.), 141
Los Angeles and Long Beach seaports, 25, 27–30, 40(fig.), 184
Los Angeles River, 29, 131
low-temperature high pressure metamorphism, 83–86

magma formation during subduction, 32, 65(fig.), 68(fig.), 70–71
magnetic directions preserved in rocks and sediments, 80–81, 163–164
magnetic reversals, 163–164
magnitudes of earthquakes: *See* earthquakes
Malibu Creek and Malibu Canyon, 43, 113, 144(fig.), 132(fig.), 139, 186–187
Malibu Point, 91(fig.), 113–115, 185
managed retreat from eroding coastlines, 123, 140, 173
marine terraces, 145–153, 179–180, 183–185
ages, 146–147, 149–153
formation, 149–153
on Palos Verdes Peninsula, 151(fig.)
on San Clemente Island, 146(fig.)
along San Diego coast, 143–145, 147(fig.)
uplift of near Ventura, 55
Matilija Dam and Matilija Reservoir, 130–131, 132(fig.), 137–140, 188
Mercalli intensity scale for earthquakes, 51
metamorphic core complexes, 82–83
metamorphic rocks, 80–88, 191–192
metamorphism, 83–86
mid-ocean ridges, 62(fig.), 63–65, 66(box), 75, 163. *See also* seafloor spreading
Milankovitch cycles, 161–165
Milankovitch, Milutin, 156(fig.), 159–161
Miyako (Japan) tsunami, 25–26
moment magnitude scale for earthquakes, 50
Monterey Canyon, 137(box)
Monterey Formation, 44, 80–81, 189, 190, 192
moraines, 154, 155
Mount Soledad (San Diego), 57, 179–180
multibeam bathymetry, 37, 38(fig.), 112(fig.)
Munk, Walter, 92(box)

North American Plate, 5–15, 18, 64(fig.), 65(fig.), 72(fig.), 74–88
capture of pieces by Pacific Plate, 74–88
movement relative to the Pacific Plate, 5–15, 17
North Pacific Gyre, 19

Northern Channel Islands, 3, 77, 87, 114, 117

ocean trenches, 26, 32, 33(fig.), 35, 64–66, 74(fig.), 75–76. *See also* subduction
Oceanside Harbor, 182
oil (petroleum) formation, 44–45
oil seeps at Carpinteria, 44, 189
ooze, 63, 70
ophiolites, 62(fig.), 67(fig.), 68(fig.)
oxygen isotopes, 150(fig.), 161–164

Pacific Flyway, 165
Pacific Plate, 5–15, 17, 26, 45, 61, 64(fig.), 65(fig.), 74–88, 171
 formation of the Basin and Range, 74(fig.)
 formation of the San Andreas fault, 74(fig.)
 movement relative to North American Plate, 5–15, 17
 tectonic capture of Southern California by, 76–88
Pacific Ring of Fire, 32–33
Pacific storm centers, 92–93
Parsons, Mike, 102, 104–105
Palos Verdes Peninsula (Palos Verdes Hills), 28, 40, 111, 112(fig.), 119, 149, 151(fig.), 184–185
Palos Verdes Slide, 37, 40–41
Pangaea, 66
Patton Escarpment, 18, 20, 91(fig.)
pillow basalt, 62–63, 69, 71, 84(fig.), 85. *See also* seafloor spreading
plankton, 44, 89, 161–164
plate tectonics, 63–88, 163(box). *See also* North American Plate, Pacific Plate, seafloor spreading, subduction
Point Conception, 17–22, 26, 40, 92–93, 105, 122
Point Fermin, 119–121, 184
Point Vicente, 151(fig.), 184–185

Queen Mary (ship), 29, 184

radioactive decay (use in geologic dating), 146–148
refraction. *See* wave refraction
Redondo Breakwater, 91(fig.), 111–113, 185
Redondo Canyon, 111–113, 138 (fig.), 185
releasing bends (in faults), 34(fig.), 35
replenishment of beaches, 140–142
restraining bends (in faults), 34(fig.), 35, 57, 58(fig.)
revetments, 123, 126, 131
Richter scale for earthquakes, 50
Rincon Point, 90, 91(fig.), 115–117, 188–189
Rindge Dam, 132(fig.), 139, 186
Ring of Fire, 32–33
rhyolite, 3
Rose Canyon fault, 57–59, 106, 179–180

San Andreas fault, 2, 6(fig.), 7–14, 15(fig.), 16(box), 18, 28, 35, 47–59, 72(fig.), 73–79
 earthquakes along, 47–53
 geologic formation of, 73–79
San Bernardino Mountains, 11
San Clemente Island, 90, 144, 146(fig.), 149, 191
San Diego Regional Beach Sand Project, 141–142
San Diego River, 58(fig.), 131
San Elijo Lagoon, 181
San Gabriel Mountains, 11n
San Gabriel River, 131
San Luis Rey River, 131
San Mateo Creek, 108–110, 182
San Mateo Point, 109–110, 182
San Miguel Island, 3–7, 14, 61, 176, 192–193
San Nicolas Island, 149
San Onofre Breccia, 86–88, 182–183
San Pedro Basin, 28–29, 40–41
San Pedro Escarpment, 28, 40–41, 112(fig.), 185
Santa Ana River, 131
Santa Barbara Harbor, 134(fig.), 135, 189–190
Santa Catalina Island. *See* Catalina Island
Santa Clara River, 131, 187
Santa Margarita River, 131
Santa Maria River, 131
Santa Monica Mountains, 3, 11, 43, 77, 87, 186–187
Santa Monica Shelf, 111, 112(fig.)
Santa Ynez Mountains, 3, 11, 43, 77, 130, 188, 190–191
Santa Ynez River, 131
schist, 83–85, 86(fig.), 87(fig.)
Scripps Canyon, 58(fig.), 106, 107(fig.), 111, 180–181
Scripps Coastal Reserve, 180
seafloor spreading, 63–66, 161, 163(box). *See also* mid-ocean ridges

sea level change, 143–170
and formation of coastal wetlands, 166–170
and formation of marine terraces, 147(fig.), 149–153
during ice ages, 149, 150(fig.)
highstands, 151–153
lowstands, 153
rise if all ice melted, 149
rise in future, 123, 167–170
rise since Last Glacial Maximum, 121–122, 150(fig.)
stability in recent millennia, 110, 121–122, 143, 173
seaports of Los Angeles and Long Beach, 25–30, 40(fig.), 184
seawalls, 121, 123, 126–131, 133, 140–142, 173
seismic risk. *See* earthquakes
Sierra Nevada, 9, 73
Sierran Plate, 9–11
slip rates of faults, 55–59
sonar, 37
Southern California Bight, 16–23, 35, 92–93, 105, 122, 133, 136(box), 173
longshore drift in, 133–135, 138(fig.)
ocean currents in, 18–20
ocean temperatures in, 21(fig.)
surfing in, 105–117
waves in, 20–23
Storm (good dog), 144(fig.)
Storm enters in the Pacific Ocean, 92–93
subduction, 32–33, 35, 62(fig.), 64(fig.), 63–87
accretionary wedge formed by, 65(fig.), 70, 71
assembly of California by, 63–73
of Farallon Plate, 73–76
magma formation during, 70–71
metamorphism caused by, 83–86
tsunamis formed by, 32
submarine canyons, 107(fig.), 112(fig.), 135–137, 138(fig.)
Sunken City, 119–121, 184
Surfer's Point (Ventura), 123
surfing, 89–117
surf science concepts, 100–101, 105
swell formation, 92–93, 95(fig.)
relation to surfing waves, 96–101
swell shadows, 21–23, 106–108, 114–117
swell windows, 21–23, 106–108, 114–117

tectonic evolution of western North America, 9–14
tectonic plates, 5–15, 149
earthquakes and, 14–15
movements in Southern California, 5–14
world map of, x–xi
terranes. *See* accreted terranes
Tijuana River Estuary, 179
time. *See* geologic time
Torrey Pines State Park and State Beach, 126(fig.), 145(fig.)
tow-surfing, 104
Transverse Ranges, 6(fig.), 8(fig.), 9–11, 35, 48, 77, 190
trenches. *See* ocean trenches, subduction
Trestles, 91(fig.), 108–111, 182
tsunamis, 25–41
behavior in deep versus shallow water, 29
from Catalina fault, 35
in the Continental Borderland, 33, 35, 36–41
fictional scenario for Southern California, 28–30
history and examples, 31, 35
inundation maps for, 172
in Japan, 25–26
in Lituya Bay (Alaska), 31, 32(fig.), 37, 41
Pacific trenches and, 33(fig.)
preparation for, 172–173
risk from far-traveled, 35–36
subduction and, 26, 32, 35,
'tidal wave' misnomer, 29
from undersea landslides, 36–41
turbidites, 73
turbidity currents, 137(box)

uniformitarianism, 69
uranium-thorium in geologic dating, 148
Urey, Harold, 162

Ventura Anticline, 187–188
Ventura Hills, 187–188
Ventura River, 43, 130, 131
volcanic arcs, 65(fig.), 70–73
von Humboldt, Alexander, 154, 155

Walker Lane–Eastern California Shear Zone, 9n, 10(fig.), 12, 52(fig.)
water recycling, 175
water supply for Southern California, 174–176
wave base, 94–101
wave breaking, 97

wave-cut platforms, 143–144. *See also* marine terraces
wave dispersion, 92(box), 95–96
wave orbits, 94–95
wave period, 94–101
importance for refraction, 97–100
relation to wave base, 97
wave physics and behavior, 90–101
wave refraction, 97–101
at Black's Beach, 106–107
over Cortes Bank, 102, 103(fig.)
at Malibu Point, 113–114
at Redondo Breakwater, 111–112
at Rincon Point, 116–117
role in forming surfing waves, 100–101
at Trestles, 110
over undersea canyons, 99(fig.)
over undersea ridges, 98(fig.)
Western Transverse Ranges Block (WTRB), 8(fig.), 77–82, 85, 190, 192
geologic rotation of, 77–82, 190
magnetic evidence of rotation, 80–81, 190
wetlands, 165–170
Wheeler Gorge, 188
Willamette Valley (Oregon), 64(fig.), 65(fig.), 70, 73
wind and ocean wave formation, 91–95
Wybenga, Matt, 104